WITHDRAWN

MATHEMATICAL PREPARATION FOR LABORATORY TECHNICIANS

JOSEPH I. ROUTH, Ph.D.
Professor, Department of Biochemistry,
College of Medicine, The University of Iowa,
Iowa City, Iowa

 SAUNDERS GOLDEN SERIES

W. B. SAUNDERS COMPANY

PHILADELPHIA • LONDON • TORONTO

1971

W. B. Saunders Company: West Washington Square
Philadelphia, Pa. 19105

12 Dyott Street
London, WC1A 1DB

833 Oxford Street
Toronto 18, Ontario

Mathematical Preparation for Laboratory Technicians   ISBN 0-7216-7735-5

© 1971 by W. B. Saunders Company. Copyright under the International Copyright Union. All rights reserved. This book is protected by copyright. No part of it may be reproduced, stored in a retrieval system, or transmitted in any form or by any means, electronic, mechanical, photocopying, recording, or otherwise, without written permission from the publisher. Made in the United States of America. Press of W. B. Saunders Company. Library of Congress catalog card number 79-151685.

Print No: 9   8   7   6   5   4   3

Preface

The majority of students have been exposed to a mathematics course every year throughout grade school, junior high, senior high and often in college. This extensive diet of arithmetic, algebra, plane and solid geometry, trigonometry and sometimes differential and integral calculus should result in a large number of mathematically oriented students. It would appear that students electing to specialize in some branch of chemistry or in a health-related science such as medical technology, pharmacy, dentistry or medicine should have achieved a firm foundation in mathematics and the ability to apply it to their specialty.

The author's contention, based on over three decades of teaching experience with students in the aforementioned disciplines, is that the multiple mathematics course system has not achieved a practical goal. It is obvious that all these courses are not required to prepare a student to add, subtract, multiply or divide properly. Algebraic, geometric and trigonometric functions are often memorized but seldom applied to practical situations. Unfortunately the meaningful application of mathematics often awaits the choice of a professional career. Long after the great mass of equations, theorems and exercises is forgotten the student is suddenly faced with the unfamiliar mathematical requirements of a scientific specialty.

At the risk of insulting the intelligence of mathematically oriented students this book stresses arithmetical common sense. In a classic sense it is not truly a mathematics book, but rather a book that reviews techniques in mathematics and applies these techniques to the solution of practical problems in a chemistry or health science laboratory.

The first three chapters set the stage for the application of mathematics in the laboratory. The material is obviously a review for many students and hopefully presents a practical foundation for techniques of problem solving in later chapters. If adequate time is spent considering the explanations and working the problems in these chapters, the subsequent material will be much easier to understand. Chapter 4 considers laboratory problems in the preparation of solutions, expressions of concentration, conversion from one type of concentration unit to another and dilution problems that commonly occur in the laboratory. The preparation of graphs, the substance of Chapter 5, often poses a difficult problem for laboratory technicians faced with calibration, standard curves, or kinetic measurements. Calculations involving spectrophotometric measurements are considered in Chapter 6, followed by the application of elementary statistics to quality control programs in Chapter 7. Chapters 8 and 9 cover the practical application of calculations involving pH and hydrogen ion concentration and buffer solutions to the preparation of buffers and the subject of acid-base balance. The final chapter is concerned with explanations of renal clearance test calculations.

To enable the student to check his understanding of the application of the material to the solution of problems and exercises in the book, answers for each exercise are given in the Appendix.

The author is indebted to so many students, instructors and laboratory technicians for their questions, problems and discussions that it would be most difficult to thank specific individuals for their role in the conception of the book. Susan Kracht deserves special thanks for her assistance in the preparation and typing of the manuscript. Grateful appreciation is due all of the instructors who criticized the manuscript prior to publication, including Sister Elizabeth Kramer, John Abadi, Carla Salmon, Jan Platt, Wayne Patton and Martha Olliver. Most important is the opportunity to express loving thanks to my wife Dorothy for her constant encouragement and insistence that the book would help present and future technicians. Finally the author is pleased to acknowledge the continual assistance of the editors and production staff of the W. B. Saunders Company.

JOSEPH I. ROUTH

Contents

Chapter 1

SIMPLE MATHEMATICAL MANIPULATIONS 1

    1.1 Handling Numbers 4
    1.2 Establishing the Answer and the Correct Decimal Point 6

Chapter 2

LOGARITHMS 9

    2.1 The Characteristic 9
    2.2 The Mantissa 10
    2.3 The Logarithm of a Number 11
    2.4 The Number Corresponding to a Given Logarithm 12
    2.5 Multiplication 13
    2.6 Division 14
    2.7 Powers of Numbers 15
    2.8 Roots 16
    2.9 Logarithm of Numbers Expressed as Powers of Ten 17

Chapter 3

THE SLIDE RULE 20

    3.1 The Division of Numbers on the Slide Rule 21
    3.2 Multiplication 23
    3.3 Setting the Decimal Point 24
    3.4 Multiplication of More Than Two Numbers 25

3.5 Division	25
3.6 Combination — Multiplication and Division Problems	27
3.7 Squares and Square Roots	29
3.8 Conversion and Proportion	31
3.9 The Circular Slide Rule	33

Chapter 4

CONCENTRATION OF SOLUTIONS — 41

4.1 Percentage Solutions	41
4.2 Specific Gravity	43
4.3 Hydrates and Water of Hydration	44
4.4 Molar Solutions	45
4.5 Molal Solution	47
4.6 Normal Solution	47
4.7 Dilution of Solutions	48
4.8 Conversion Problems	51

Chapter 5

THE PREPARATION OF GRAPHS — 55

5.1 The Choice of Units for the Coordinates	56
5.2 The Glucose Tolerance Curve	58
5.3 The Preparation of Standard Curves	58

Chapter 6

SPECTROPHOTOMETRIC CALCULATIONS — 65

Chapter 7

QUALITY CONTROL STATISTICS — 71

Chapter 8

HYDROGEN ION CONCENTRATION AND pH — 76

CONTENTS

Chapter 9
BUFFERS **80**

    9.1 Acid-Base Buffer Calculations 84

Chapter 10
RENAL CLEARANCE TEST CALCULATIONS **89**

    10.1 Urea Clearance Test 89
    10.2 Maximum Clearance 89
    10.3 Standard Clearance 90
    10.4 Creatinine Clearance 91

APPENDIX **96**

ANSWERS TO EXERCISES **100**

INDEX **109**

CHAPTER 1

SIMPLE MATHEMATICAL MANIPULATIONS

Even before you finish high school most of you have achieved confidence in your ability to add, subtract, multiply and divide. If you make a mistake in one of these operations, it is usually the result of carelessness. Yet many of you do not always know when to multiply and when to divide or the right numbers to use in a particular problem unless you have memorized a formula or equation. Equally apparent is your lack of confidence in the correctness of a specific answer you obtain in a calculation. Often a large share of your time spent in learning mathematics has centered around memorization of equations. When you leave the classroom for any length of time, the equations fade from your memory or become distorted or modified in a way that always produces incorrect answers. The inability to accept a specific figure for an answer or difficulty in establishing the exact magnitude of the value also stems from excessive emphasis on memorization. If a fraction of your mental effort had been spent on **arithmetical common sense**, your laboratory problems would seem less formidable. With the mathematical background possessed by most students it is not too difficult to achieve arithmetical common sense. This is the ability to reason out a problem, to make sense out of the values used in multiplication and division and most of all to obtain a correct answer and to know that it is correct. Many people marvel at the ability of a person to figure out problems in his head. There is obviously nothing magical involved but merely an orderly process of making sense out of the manipulations required to solve a problem and occasionally being able to mentally calculate the exact values. This process is not essential to arithmetical common sense but is often quite dramatic. It is sufficient to be able to think through the problem in terms of rounded off values and come up with a sensible approximate answer.

Let us start with some very simple examples and work our way into laboratory mathematics. Problems involving the English system, rather than the metric system, of measurement may serve as a good start. If someone asked you how many feet there were in a yard or how many inches in a foot, you would automatically remember 3 feet to the yard and 12 inches to the foot. You may not remember how many yards there are in a mile, but if you were told that there were 5280 feet in a mile, you could probably work out in your head the number of yards per mile. Your mental gymnastics might operate as follows:

> I know there are 3 feet in a yard, so 5280 divided by 3 would equal 1 with 2 to carry, and 3 goes into 22 seven times with 1 to carry, and 18 divided by 3 is 6; so if I remember correctly, the answer would be 1760 yards in a mile.

It would not make sense to estimate 176 as an answer since 3 times 176 would not come close to 5280. This is a simple example of working a problem in your head and making sense out of the answer.

The unit of weight in the metric system is the kilogram, although 1/1000 of a kilogram, called a gram, is a very common measure of weight. In the laboratory many methods involve milligrams, which are 1/1000 of a gram, or micrograms, 1/1000 of a milligram. The conversion of these units is not too difficult to master mentally. Yet in the supermarkets many items are sold by the pound, and even people are still commonly weighed in pounds. If you are told that a kilogram equals approximately 2.2 pounds, how could you use this value to convert pounds to kilograms, pounds to grams and kilograms to pounds? Again, you reason that it takes over 2 lbs to make 1 kg, so a 150 pound person is going to weigh less than one-half that value in kilograms, or less than 75 kg. In fact, 150 divided by 2.2 is about 68 kg. It still sounds more interesting to hear that a beauty queen tips the scales at 110 lbs rather than 50 kg. Chemicals are often sold in 1 lb or 5 lb bottles, and we use the chemicals in grams. From our original approximation of 1 kg equals 2.2 lbs we can state that 1000 gm equals 2.2 lbs or 1 lb equals approximately 450 gm.

In conversion problems it is sometimes difficult to avoid the use of memorized formulas or equations. This becomes clear if we examine the common centigrade and Fahrenheit temperature scales. The centigrade scale is often referred to as the Celsius scale, and degrees on the centigrade scale are termed degrees Celsius. We will use degrees C (°C) as degrees centigrade since this is relevant to the explanation of the basis of the scale that follows. The centigrade scale was designed to make the freezing point of water read 0°C and the boiling point read 100°C, giving rise to the name "centigrade." The Fahrenheit scale reads 32°F at the freezing point and 212°F at the boiling point of water. Between the freezing and boiling points of water there are 100 centigrade and 180 Fahrenheit intervals or degrees; therefore, 100 centigrade intervals equal 180 Fahrenheit intervals. From this we conclude that

$$1° \text{ centigrade} = \frac{180}{100} = \frac{9°}{5} \text{ Fahrenheit}$$

SIMPLE MATHEMATICAL MANIPULATIONS

and

$$1° \text{ Fahrenheit} = \frac{100}{180} = \frac{5°}{9} \text{ centigrade}$$

Even though the freezing point of water is 0° centigrade and 32° Fahrenheit, the temperature reads the same on both scales at −40°. The simplest equations for the conversion of °C to °F or °F to °C are based on the 9/5 and 5/9 relationship and the value 40. In either case add 40 to the original temperature, multiply by either 5/9 or 9/5 and subtract 40 from the result. To decide which fraction to use it helps to choose a common point like the boiling point of water, 212°F and 100°C. It can readily be seen that the Fahrenheit value is higher than the centigrade value; therefore you would use the largest fraction, 9/5, to convert centigrade to Fahrenheit. The following examples may help illustrate the point.

Examples

To convert 100°C to Fahrenheit add 40, 100 + 40 = 140.
Multiply by $\frac{9}{5}$, 140 × $\frac{9}{5}$ = 252.
Subtract 40, 252 − 40 = 212°F.

Or in outline form to convert 32°F to degrees centigrade:

32°F + 40 = 72
72 × $\frac{5}{9}$ = $\frac{360}{9}$ = 40
40 − 40 = 0°C

Conversion of values below zero requires a sensible approach to the algebraic sum of the signs. To convert −15°F to degrees centigrade:

−15°F + 40 = +25
+25 × $\frac{5}{9}$ = $\frac{125}{9}$ = +14
+14 − 40 = −26°C

CHAPTER 1

1.1 HANDLING NUMBERS

Numbers are very useful; in fact, they are essential in science courses and in the laboratory. A knowledge of the simplest way to handle numbers is obviously an asset in calculations. In general we should always express numbers in a form that can be readily used and equally readily understood. A uniform practice having many advantages is to express a large or small number as an integer (a number between 1 and 10) multiplied by a power of ten. As examples: $1000 = 1.0 \times 10^3$; $50,000 = 5 \times 10^4$; $0.001 = 10^{-3}$ and $0.000062 = 6.2 \times 10^{-5}$. If you count the number of zeros required to move a decimal point to obtain a number between 1 and 10 in the first two examples, you will obtain the correct positive power of ten.

$$1.000 = 1 \times 10^3 \qquad 5.0000 = 5 \times 10^4$$
$$321 4321$$

The same scheme may be applied to large, more complex numbers such as 135,000 and 2467.

$$1.35000 = 1.35 \times 10^5 \qquad 2.467 = 2.467 \times 10^3$$
$$54321 321$$

If you count the number of zeros and the first whole number, you have to move the decimal point to obtain a number between 1 and 10 in the following two examples; the correct negative power of ten is thus obtained.

Examples

$$0.001 = 1 \times 10^{-3} \qquad 0.000062 = 6.2 \times 10^{-5}$$
$$123 \phantom{= 1 \times 10^{-3} \qquad 0.000}12345$$

Further examples of negative powers of ten are:

$$0.000146 = 1.46 \times 10^{-4} \qquad 0.000000734 = 7.34 \times 10^{-7}$$
$$1234 \phantom{= 1.46 \times 10^{-4} \qquad 0.000000}1234567$$

Since, as will be seen in Chapters 2 and 3, the exponent of ten is identical with the characteristic of the corresponding logarithm of a number, the use of logarithms or a slide rule is simplified.

To multiply numbers expressed as powers of ten, first multiply the numbers (between 1 and 10) that are on the left of 10 to the power, then add the powers of ten.

SIMPLE MATHEMATICAL MANIPULATIONS

$$200{,}000 \times 3{,}000{,}000 = 2 \times 10^5 \times 3 \times 10^6 = 2 \times 3\,(10^{5+6}) = 6 \times 10^{11}$$

The same method may be used for more complex numbers and calculations.

Example

$$2{,}230{,}000 \times 56{,}700 = 2.23 \times 10^6 \times 5.67 \times 10^4$$

or

$$2.23 \times 5.67\,(10^{6+4}) = 12.64 \times 10^{10} \text{ or } 1.264 \times 10^{11}$$

To multiply numbers expressed as negative powers of ten, first multiply the numbers (between 1 and 10) that are on the left of 10 to the minus power, then add the negative powers of ten.

$$0.000002 \times 0.0000003 = 2 \times 10^{-6} \times 3 \times 10^{-7} = 2 \times 3\,(10^{(-6)+(-7)}) = 6 \times 10^{-13}$$

The same scheme is followed to divide numbers expressed as powers of ten or negative powers of ten.

Example

$$\frac{600{,}000}{2000} = \frac{6 \times 10^5}{2 \times 10^3} = \frac{6\,(10^{5-3})}{2} = 3 \times 10^2 = 300$$

As seen in the above example, first divide the numbers (between 1 and 10) that are on the left of the powers or negative powers of ten, then subtract the power of ten in the denominator from that in the numerator. The following is an example of division of numbers expressed as negative powers of ten.

Example

$$\frac{0.0008}{0.00002} = \frac{8 \times 10^{-4}}{2 \times 10^{-5}} = \frac{8}{2}(10^{(-4)-(-5)}) = 4 \times 10^1 = 40$$

Other powers and roots of numbers other than 10 are often used, but commonly only squares, cubes, square roots and cube roots of numbers are encountered. Logarithms are conveniently used to carry out complex calculations and those involving squares, cubes, square roots and cube roots, as will be seen in the next chapter.

1.2 ESTABLISHING THE ANSWER AND THE CORRECT DECIMAL POINT

A major problem in laboratory calculations is the proper setting of the decimal point in the final calculated value. An example of the change of volume of a gas with a change in pressure and temperature may lead to the following calculation:

$$V = 625 \times \frac{810 \times 283}{730 \times 353}$$

An inspection of the two fractions 810/730 and 283/353 reveals that they will about balance each other and yield a value close to 1 to be multiplied times 625, so the answer should approximate 625. Actually it is 555, but the decimal is correct and the answer makes sense. A calculation whose answer has a decimal point that may be more difficult to set would be as follows:

$$\frac{95.62 \times 1.211 \times 6.78 \times 3.45}{22.31 \times 3.1416 \times 9.965}$$

In this type of problem we simplify the approximate solution by rounding off values and judiciously give and take on the values. You might reason as follows:

$$\frac{100 \times 1 \times 10 \times 2}{20 \times 3 \times 10}$$

Mentally $100 \times 10 = 1000 \times 2 = 2000$ divided by $20 \times 3 = 60 \times 10 = 600$ and 2000/600 is the same as 20/6, or approximately 3.3; the correct answer is close to 3.88.

Other calculations may not result in such a simple setting of the decimal point in the answer. The knowledge gained in Section 1.1 may be used to establish the correct answer and decimal point in more complex calculations:

$$\frac{200{,}000 \times 3000 \times 0.0004}{500 \times 0.0012 \times 10{,}000} = \frac{2 \times 10^5 \times 3 \times 10^3 \times 4 \times 10^{-4}}{5 \times 10^2 \times 1.2 \times 10^{-3} \times 1 \times 10^4}$$

SIMPLE MATHEMATICAL MANIPULATIONS

Since to multiply we add the powers of ten, the fraction would be expressed as:

$$\frac{2 \times 3 \times 4}{5 \times 1.2 \times 1} \times \frac{10^{5+3+(-4)}}{10^{2+(-3)+4}} = 4 \times \frac{10^4}{10^3}$$

To divide we subtract the power of ten in the denominator from that in the numerator, or $4 \times 10^{4-3} = 4 \times 10 = 40$. Manipulation of a series of more complex numbers may be illustrated as follows:

$$\frac{1232 \times 3.141 \times 627.6 \times 72.1}{29.42 \times 411.2 \times 6.95}$$

As powers of ten:

$$\frac{1.232 \times 10^3 \times 3.141 \times 6.276 \times 10^2 \times 7.21 \times 10^1}{2.942 \times 10^1 \times 4.112 \times 10^2 \times 6.95}$$

$$= \frac{1.232 \times 3.141 \times 6.276 \times 7.21}{2.942 \times 4.112 \times 6.95} \times \frac{10^{3+2}}{10^2}$$

$$= \text{approx.} \frac{1 \times 3 \times 6 \times 7}{3 \times 4 \times 7} = \frac{126}{84} \times 10^{5-2} = 1.8 \times 10^3$$

$$= \text{by actual calculation } 2.08 \times 10^3 = 2080$$

In laboratory calculations it is often of great assistance to know the range of values expected from a patient and thus from the method and the proper calculation of values by the use of that method. There are many instances in which the method and the calculation are properly designed to yield, for example, values in a range of 5 to 20 mg/100 ml. It is always somewhat of a shock when a laboratory technician insists that the patient's blood level is 143 mg/100 ml because he obtained that answer from his calculation. The use of common sense in this calculation implies not only knowing how to establish the correct decimal point but also recognizing the range of levels you would ordinarily expect. Before insisting on 143 mg/100 ml or even entering it in the patient's record, it would be wise to recalculate the problem.

EXERCISES

1. If peaches were priced at $0.77 a kilogram, what would they cost per pound?

2. Some professional football players can command an annual salary equivalent to their weight in gold. If gold sells for $1000 a kilogram and a player weighs 220 pounds, what is his annual salary?

3. A technician's child appeared to be running a high fever. The only thermometer available was calibrated in degrees centigrade and gave a reading of 40°C. What was the child's temperature in degrees Fahrenheit?

4. Dry ice and acetone mixtures are often used for the rapid freezing of biological specimens. A temperature of -90°C may be obtained in this fashion. Express this in degrees Fahrenheit.

5. Room temperature is usually about 74°F. What would be the reading on the centigrade scale?

6. In the following examples give the approximate answer obtained by mental arithmetic to aid in setting the correct decimal point.

a. $\dfrac{1152.6 \times 3.1416 \times 809.61}{38.46 \times 12.124 \times 4.891}$

b. $633 \times \dfrac{820}{760} \times \dfrac{290}{353}$

c. $\dfrac{24.782 \times 2.173 \times 111.22}{42.36 \times 9.77 \times 2.676}$

7. Express the following numbers as powers of ten or negative powers of ten.

a. 2,200,000

b. 46,000

c. 151,000

d. 0.0003

e. 0.00000044

f. 0.000000005

8. Carry out the following calculations after first converting to powers or negative powers of ten.

a. 410000 × 23000

b. 5600000 × 720000

c. 0.000031 × 0.00022

d. $\dfrac{8100000}{300000}$

CHAPTER 2

LOGARITHMS

The logarithm of a number is the power to which a given base must be raised to equal that number. In the common system of logarithms the base is always 10. For example, $1000 = 10^3$, so the logarithm of 1000 to the base 10 is 3. Similarly 2 is the logarithm of 100 because $10^2 = 100$. Also 0.699 is the logarithm of 5 since $10^{0.699} = 5$. We have just seen that small fractions or numbers less than one (1.0) can be expressed as 10 to a minus power. For example, $0.001 = 10^{-3}$ and the logarithm of 0.001 would be -3. The figure 0.1 can be expressed as 10^{-1} and thus has a logarithm of -1. It should not be too difficult to decide that the log of 1 = 0 and that the log of 10 = 1. Most logarithms are not as simple as the above examples and consist of two major parts, the **characteristic** and **mantissa**. The logarithm of 752, for example, is 2.8762; in this example the number to the left of the decimal point, 2, is called the characteristic, and the decimal part, 0.8762, is called the mantissa.

2.1 THE CHARACTERISTIC

The characteristic of the logarithm of a number equal to or greater than one is one less than the number of digits to the left of the decimal point. The characteristic of the log of 752 would then be 2, while that of 75.2 would be 1 and the characteristic of 7.52 would be 0.

Examples

The characteristic of 6271 = 3
1.5 = 0
30.2 = 1
512 = 2

The characteristic of the logarithm of a number less than one is negative and is one more than the number of zeros after the decimal point. A negative characteristic may be expressed by placing a negative sign over the number, as $\bar{1}$, $\bar{2}$ or $\bar{3}$, or as a positive number minus 10; that is $\bar{1}$ would be 9 - 10, $\bar{2}$ would be 8 - 10 and $\bar{3}$ would be expressed as 7 - 10. The use of a negative characteristic with a positive mantissa in the log of 0.005 = $\bar{3}$, or the subtraction of the 3 from 10 - 10 to give 7 - 10, is employed in most calculations involving logarithms.

Examples

The characteristic of 0.005 = $\bar{3}$ or 7 - 10
0.10 = $\bar{1}$ or 9 - 10
0.025 = $\bar{2}$ or 8 - 10

2.2 THE MANTISSA

The mantissa is the decimal part of the logarithm and is independent of the decimal point in the original number. The mantissa of the log of a number is found in logarithm tables, and although the decimal points are not printed in the table, each mantissa is understood to have a decimal point in front of it. The L scale of a slide rule may also be used to obtain the mantissa. This procedure will be explained in the following section. The logarithm table (see the Appendix at the end of this book) contains mantissas of numbers from 1 to 9999 to four decimal places. This is called a four-place table and has the first two digits of the number in the left column under N with the third in a row across the top from 0 to 9. All of the numbers in the body of the table from 0000 to 9996 are mantissas. The mantissa of a two digit number is found under the 0 column opposite the number; for example, the mantissa of 65 is 0.8129. For a three digit number the mantissa is found under the column corresponding to the third digit opposite the first two digits; thus the mantissa of 455 is 0.6580.

Examples

The mantissa of 5.0 = 0.6990
56 = 0.7482
820 = 0.9138
82 = 0.9138
8.2 = 0.9138

LOGARITHMS

2.3 THE LOGARITHM OF A NUMBER

To find the complete logarithm of a number we first establish the characteristic and follow it with the mantissa as the decimal part found in logarithm tables. A four-place table yields mantissas of three digit numbers directly and four digit numbers by interpolation between the numbers across the top of the table. The use of the four-place table will be illustrated in the following examples.

Examples

Find the log of 672. The characteristic is 2, and the mantissa from the tables is 0.8274, so the complete logarithm of 672 is 2.8274. Finding the log of 6725 involves an additional step. The characteristic is 3, and the mantissa is determined as follows:

$$
\begin{aligned}
\text{The mantissa of } 6720 &= 0.8274 \\
6730 &= 0.8280 \\
6725 &= 0.8277
\end{aligned}
$$

$$
\begin{array}{r}
0.8280 \\
-0.8274 \\
\hline
0.0006 \\
0.5 \\
\hline
0.0003
\end{array}
$$

The reasoning would be as follows. 6725 is half of the way (0.5) between 6720 and 6730, whose mantissas are 0.8274 and 0.8280, respectively, with a difference of 0.0006; 0.5 of 0.0006 is 0.0003. Adding this to 0.8274 = 0.8277, the mantissa of 6725. The complete logarithm would then be 3.8277. You will notice that the logarithm table has a set of proportional parts to assist you in interpolation between the last two digits. In any horizontal row in the table the difference between the 10 mantissas is nearly the same number, and the number across the top of the proportional parts set are tenths of the difference between any two mantissas in that row. A satisfactory logarithm of 6725 could be obtained by following the characteristic of 3 with the mantissa obtained by reading the 672 from the table as 0.8274 and adding the proportional part in that row under the 5 on the top, giving 0.0003 + 0.8274 = 0.8277, or a log of 3.8277. A little practice with examples with establish this principle in your mind.

EXERCISES

Find the logarithm of

A. 5.5   = _____

B. 431   = _____

C. 0.062 = _____

D. 3244   = _____

E. 0.15   = _____

2.4 THE NUMBER CORRESPONDING TO A GIVEN LOGARITHM

To find the number that corresponds to a particular logarithm we have to reverse the process of finding the logarithm. This may be done with a slide rule (as will be explained later), by use of a table of **antilogarithms** (reverse logarithms may be a better name) or by finding the digits in the number that correspond to the mantissa of the logarithm in a regular log table and setting the decimal point from the characteristics of the logarithm. As an example let us find the number whose logarithm is 2.4843. The mantissa 0.4843 is located in the row whose first two digits are 30 and is under the third digit 5 in the top row. The three digits are therefore 305 and with a characteristic of 2 would have three digits to the left of the decimal point, resulting in the number 305. We can then say the log of 305 is 2.4843, or the reverse logarithm, or antilogarithm, of 2.4843 is 305. The same example can be solved directly by use of the table of antilogarithms. In this table the first two digits of the mantissa are given as decimals in the left column, while the third digit is shown across the top and the fourth digit of the mantissa is found by use of the proportional parts. The number corresponding to the logarithm is given in the body of the table from 1000 to 9977. To obtain the digits of the number from the mantissa 0.4843 we start with the row having 0.48 on the left side then move over to the number 3048 under the 4 column and add 2 as the proportional part under 3, giving the four digits 3050. Again, with a characteristic of 2 the number corresponding to a log of 2.4843 would be 305.0.

LOGARITHMS

EXERCISES

Find the antilog of

A. $1.3177 =$ _____

B. $3.1805 =$ _____

C. $0.2664 =$ _____

D. $\bar{2}.5768$ or $8.5768 - 10 =$ _____

E. $\bar{1}.3667$ or $9.3667 - 10 =$ _____

2.5 MULTIPLICATION

Calculations with the aid of logarithms follow the laws of exponents, since logarithms are actually exponents. **The logarithm of the product of two numbers is the sum of their logarithms.**

$$\log ab = \log a + \log b$$

To multiply numbers, add their logarithms and find the number that corresponds to that logarithm (the antilogarithm).

Examples

$$12.6 \times 4.3 = 54.18$$

$$\begin{array}{ll} \log 12.6 = & 1.1004 \\ \log 4.3 = & \underline{0.6335} \\ \text{Sum} = & 1.7339 \\ \text{Antilog} = 54.18 \end{array}$$

$$3.2 \times 0.063 \times 15.6 = 3.145$$

$$\begin{array}{lll} \log 3.2 & = & 0.5051 \\ \log 0.063 & = & 8.7993 - 10 \\ \log 15.6 & = & \underline{1.1931} \\ \text{Sum} & = & 10.4975 - 10 \text{ or } 0.4975 \\ \text{Antilog} & = & 3.145 \end{array}$$

EXERCISES

A. 1.35 × 14.6 = _____

B. 32.6 × 0.57 × 17.1 = _____

C. 0.004 × 0.128 × 6.3 = _____

2.6 DIVISION

From the properties of exponents, **the logarithm of the quotient of two numbers is equal to the logarithm of the numerator minus the logarithm of the denominator.**

$$\log \frac{a}{b} = \log a - \log b$$

To divide numbers, subtract the logarithm of the denominator from the logarithm of the numerator and find the number that corresponds to that logarithm (antilogarithm).

Examples

$$\frac{205.1}{62.6} = 3.27$$

$$\begin{aligned}
\log 205.1 &= 2.3120 \\
\log 62.6 &= \underline{1.7966} \\
\text{Difference} &= 0.5154 \\
\text{Antilog} &= 3.276
\end{aligned}$$

Another example combining multiplication and division

$$\frac{35.6 \times 0.72 \times 14.9}{11.3 \times 0.082} = 412.1$$

$$\begin{aligned}
\log 35.6 &= 1.5514 \\
\log 0.72 &= 9.8573 - 10 \\
\log 14.9 &= \underline{1.1732} \\
\text{Sum} &= 12.5819 - 10
\end{aligned}$$

$$\begin{aligned}
\log 11.3 &= 1.0531 \\
\log 0.082 &= 8.9138 - 10 \\
\text{Sum} &= 9.9669 - 10
\end{aligned}$$

To divide: $12.5819 - 10$
$\phantom{\text{To divide: }}\underline{9.9669 - 10}$
$\phantom{\text{To divide: 12.}}2.6150$

Antilog $= 412.1$

EXERCISES

A. $\dfrac{4.15}{0.86} =$ _____

B. $\dfrac{15.2 \times 12.6}{5.76} =$ _____

C. $\dfrac{21.1 \times 4.3 \times 622.2}{3.76 \times 29.8} =$ _____

2.7 POWERS OF NUMBERS

The logarithm of a number to the nth power is n times the logarithm of the number.

$$\log a^n = n \log a$$

To raise a number to a power, multiply the logarithm of the number by the power and find the number that corresponds to the logarithm (antilogarithm).

Examples

$$4.5^3 \text{ or } \log (4.5)^3 = 3 \times \log 4.5 = 1.9596$$

$\log 4.5 = 0.6532$
$\underline{\times\, 3}$
1.9596

Antilog $= 91.1 = 4.5^3$

$(14.23)^2$ or $\log(14.23)^2 = 2 \times \log 14.23 = 2.3064$

$$\begin{aligned}\log 14.23 &= 1.1532 \\ &\underline{\times\ 2} \\ &2.3064 \\ \text{Antilog} &= 202.5 = 14.23^2\end{aligned}$$

2.8 ROOTS

The logarithm of a number to the nth root is the logarithm of the number divided by n.

$$\log \sqrt[n]{a} = \frac{1}{n}\log a = \frac{\log a}{n}$$

To determine the nth root of a number, divide the logarithm of the number by the root and find the number that corresponds to the divided logarithm (antilogarithm).

Examples

$$\sqrt[3]{32.2} \text{ or } \log\sqrt[3]{32.2} = \frac{1}{3}\log 32.2 = 0.5026$$

$$\log 32.2 = \frac{1.5079}{3} = 0.5026$$
$$\text{Antilog} = 3.181 = \sqrt[3]{32.2}$$

$$\sqrt{1026} \text{ or } \log\sqrt{1026} = \frac{1}{2}\log 1026 = 1.5055$$

$$\log 1026 = \frac{3.0111}{2} = 1.5055$$
$$\text{Antilog} = 32.03 = \sqrt{1026}$$

LOGARITHMS

2.9 LOGARITHM OF NUMBERS EXPRESSED AS POWERS OF TEN

In chemical calculations, concentrations are often expressed as powers of ten. For example, we may use 5.6×10^8 moles of a substance or a hydrogen ion concentration of 2.5×10^{-4}. The general expression would be $a \times 10^b$ where the logarithm equals:

$$\log a + \log 10^b = b + \log a$$

Examples

$$\log 5.6 \times 10^8 = \log 5.6 + \log 10^8 = 8 + \log 5.6$$

$$\log 5.6 = 0.7482 \text{ and } 8 + 0.7482 = 8.7482$$
$$\text{Therefore } \log 5.6 \times 10^8 = 8.7482$$

To check antilog = 560,000,000 or 5.6×10^8

$$\log 2.5 \times 10^{-4} = \log 2.5 + \log 10^{-4} = -4 + \log 2.5$$

$$\log 2.5 = 0.3979 \text{ and } -4 + 0.3979 = \overline{4}.3979 = 6.3979 - 10$$

To check antilog = $0.00025 = 2.5 \times 10^{-4}$

EXERCISES

A. 776^2 = _____

B. 35.1^3 = _____

C. $\sqrt{86.2}$ = _____

D. $\log 7.2 \times 10^6$ = _____

E. $\log 5.2 \times 10^{-3}$ = _____

ADDITIONAL EXERCISES

1. Give the characteristic of the logarithm of the following:

   a. 423

   b. 0.032

   c. 0.00005

   d. 23.5

   e. 1.7

2. Give the mantissa of the logarithm of the following:

   a. 4.6

   b. 460

   c. 78

   d. 0.143

   e. 0.007

3. What is the logarithm of the following numbers?

   a. 672

   b. 1.5

   c. 34000

   d. 0.0034

LOGARITHMS

4. What is the antilogarithm of the following logarithms?

    a. 2.7810

    b. 0.5988

    c. 8.4440 - 10

    d. 1.9877

    e. 7.6314 - 10

5. Carry out the following calculations using logarithms.

    a. 35 × 66

    b. 1.68 × 5600

    c. 3.1416 × 620 × 15

    d. $\dfrac{3200}{27}$

    e. $\dfrac{0.002 \times 28.5 \times 143.5}{2.47 \times 3.1416}$

6. Work the following problems with logarithms.

    a. 17.56^3

    b. 1.46^2

    c. $\sqrt[3]{42.3}$

    d. $\log 8.1 \times 10^5$

    e. $\log 2.4 \times 10^{-6}$

CHAPTER 3

THE SLIDE RULE

A slide rule is a simple instrument containing several scales on which **whole numbers are spaced at intervals proportional to their logarithms**. This arrangement permits rapid mechanical calculations by the use of logarithms and antilogarithms. **Multiplications**, for example, are carried out by **adding logarithms** of numbers and reading the antilogarithm of the answer. **Divisions** are readily carried out by the opposite manipulation of **subtracting logarithms** and reading the antilogarithm of the answer on a slide rule scale.

A careful examination of an inexpensive slide rule will reveal three major parts, the body, the slide and the indicator with a hairline. The **body** usually has scales such as the A, D, K, and L scales. The **slide** has projections on the sides that are closely fitted into grooves on the body of the slide rule for smooth back-and-forth movement. The slide ordinarily has B, C and C_1 scales plus special scales for various calculations. The **indicator** consists of a window with a vertical hairline; the indicator slides smoothly in grooves on the outer edges of the body of the slide rule.

A comparison of the C and D scales with the L scale will reveal the basis for multiplication and division by use of logarithms on the slide rule. The L scale is linear with equal spaces between the 10 major numbers; these divisions correspond to the mantissa of logarithms found in a log table. The C and D scales also contain 10 major divisions, but the whole numbers are spaced at distances proportional to their logarithms. For example, the log of 2 is 0.301, so the 2 is located over the 0.301 on the linear or log scale L. In like manner the log of 5 is 0.699, so the 5 is located over the 0.699 on the L scale. Also, since the log of 10 is 1.000, the number 10 on the extreme right of the C or D scales is over the 1.000 on the L scale.

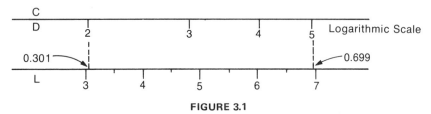

FIGURE 3.1

THE SLIDE RULE

3.1 THE DIVISION OF NUMBERS ON THE SLIDE RULE

A major problem for students first learning to use a slide rule is the division of whole numbers on the C and D scales. To correctly read the scales on a slide rule you must be thoroughly familiar with the major and minor divisions of each whole number. If we first consider the whole number divisions on the C and D scales, they will appear as in Figure 3.2.

```
D
or  1           2           3       4   5  6 7 8 9 1
C
```

FIGURE 3.2

The space between each whole number is first divided into tenths, making it relatively simple to read the first two digits of a given number. Further division of the tenths between each whole number varies with the size of the space between the numbers, as seen in Figure 3.2. The large space between 1 and 2 permits each tenth to be divided into tenths, for a total of 100 divisions, allowing a direct reading of the first three digits with the fourth digit to be estimated. An example of a number setting in this region would be 166 or 1660; move the hairline of the indicator to the right of the 1 on the D scale to the 6 (6 tenths mark) and then to the 6th mark between 6 and 7 (6 hundredth mark). This setting of 166 could represent 1660, 0.166, 16.6 or any other combination of 166. If the value was 1665, you would further move the hairline to the right halfway between the 6 and 7th hundredth mark.

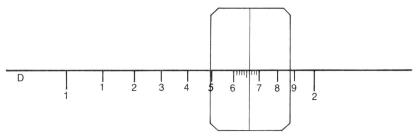

FIGURE 3.3

The area between whole numbers 2 and 4 is first divided into tenths (these divisions are not marked with numbers as they are between 1 and 2). The smaller space permits only further divisions of fifths between the tenths making a total of 50 divisions between 2 and 3 and between 3 and 4. The smallest divisions would thus represent 2 hundredths instead of 1 hundredth as between 1 and 2. A setting of 366 would be achieved by moving the hairline past 3 to the right to the 6 tenth mark and then to the third division (6 hundredths) between the 6th and 7th tenth. To set 367 you would further move the hairline halfway between third and fourth division (6th and 8th hundredths) between the 6th and 7th tenth.

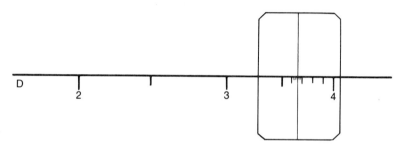

FIGURE 3.4

From whole numbers 4 to 10 each number is again divided into tenths, and each tenth is divided into halves, making the smallest divisions 5 hundredths. There would thus be 20 divisions between each whole number from 4 to 10. To set the number 665 you would move the hairline past 6 to the right to the 6th tenth and then to the mark halfway between the 6th and 7th tenth. In this region of the D scale the third digit of a number must be estimated by the spacing between the tenths if it is any number other than 5.

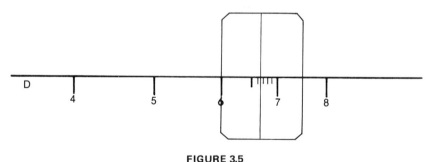

FIGURE 3.5

Before attempting any calculations with the slide rule, practice setting numbers on the D scale with the hairline. For example, locate the following numbers and check your setting with the log values from the L scale.

THE SLIDE RULE

Examples

Set 126 (corresponds to 0.100 on L scale)
3.16 (corresponds to 0.500 on L scale)
1675 (corresponds to 0.224 on L scale)
45.5 (corresponds to 0.658 on L scale)
867 (corresponds to 0.938 on L scale)
6.31 (corresponds to 0.800 on L scale)
1333 (corresponds to 0.125 on L scale)

3.2 MULTIPLICATION

Since the C and D scale numbers correspond to the mantissa of the logarithms on the L scale, the process of multiplication on the slide rule consists of adding the logarithm of two numbers and finding the number corresponding to the sum of the logarithms on the D scale. A simple example would be to multiply 2 × 2. First set the 1 on the left of the C scale over the 2 on the D scale; then move the hairline to the 2 on the C scale and read the answer below on the D scale. The answer is obviously 4, but this is also an easy one to understand — addition of the log of 2 = 0.301 (from 0 to 0.301 on the L scale) and the log of 2, or

$$0.301 + 0.301 = 0.602$$

on the L scale, which corresponds to the number or antilog of 4.

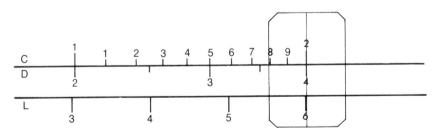

FIGURE 3.6

In general you use the 1 on the C scale at either the left or right end of the slide rule, depending on the number you are multiplying. For example, you wish to multiply 4 × 5. You will find that if you set the 1 on the left of the C scale over the 4 on the D scale, the 5 on the C scale projects out into space. By setting the

1 on the right side of the C scale over the 4 on the D scale, the hairline can be moved to the 5 on the C scale and the answer 20 found below the line on the D scale. To try a more difficult example multiply 185 × 4.46. Set the 1 on the left of the C scale over 185 and move the hairline to 4, then to 4 tenths and then to 6 hundredths, or 446, and find the answer 825 below on the D scale. Practice multiplication by working the following problems.

EXERCISES

A. 15 × 39 = _____

B. 21 × 35 = _____

C. 92 × 6.85 = _____

D. 52 × 2.5 = _____

3.3 SETTING THE DECIMAL POINT

In both multiplication and division problems on the slide rule there is a rather complex method of setting the decimal point of an answer that involves the span of a number, the sum or difference between spans and the addition or subtraction of 1 from the sum or difference of the spans, depending on which side of the slide rule the C slide projects. Unfortunately a wrong answer could result from the improper use of the span method even after the correct slide rule manipulation was carried out. **The common sense approach explained in Section 1.2 will assist you in setting the correct decimal point in your answer.** The procedure of multiplying two numbers such as 1.95 × 33.6 would begin with setting the 1 on the left of the C scale over the 195 on the D scale and moving the hairline to the 336 on the C scale; the answer is 656. The problem is the decimal point; should the correct answer be 6.56, 65.6, 656? By rounding off the figures 1.95 to 2 and 33.6 to 30 the answer would be approximately 60; 65.6 is therefore correct. In every problem in which the answer is not readily apparent the decimal point setting should be determined by rounding off the numbers and obtaining an approximate answer. As stated earlier, this process can usually be worked out mentally and is worthy of practice.

THE SLIDE RULE

3.4 MULTIPLICATION OF MORE THAN TWO NUMBERS

Laboratory calculations often involve the multiplication of three or more factors, a process which can readily be carried out on the slide rule. Take as an example the problem 3.14 × 52.6 × 0.935. First, set the 1 on the right of the C scale over 3.14 on the D scale, then move the hairline to 52.6 on the C scale. Instead of reading the answer as 165.1 on the D scale move the 1 on the right of the C scale to the hairline setting (165.1) and then move the hairline to 935 on the C scale and read the answer 1528 below on the D scale. You should be able to set the decimal by inspection since 0.935 is almost 1 and 3 × 50 equals 150 so the correct answer would be 152.8. With the hairline on the setting 1528 on the D scale you could complete the four number problem 3.14 × 52.6 × 0.935 × 75 by moving the 1 on the right of the C scale to the hairline setting (1528) and again moving the hairline to the 75 on the C scale reading the answer 1147 below on the C scale. To set the decimal point multiply 150 × 70 mentally, obtaining 10500; so the correct answer would be 11,470. Do the following exercises to gain proficiency in multiplication and the setting of the decimal point.

EXERCISES

Approximate answer

A. 5.2 × 0.876 × 12.3   =   5 × 1 × 10 = 50

B. 94.3 × 18.1 × 31.2   =   100 × 20 × 30 = 60,000

C. 7.8 × 5.6 × 19.2 × 62.3 =   10 × 5 × 20 × 60 = 60,000

D. 46.4 × 0.082 × 3.33   =   50 × 0.1 × 3 = 15

E. 110.5 × 2.56 × 0.0094   =   100 × 3 × 0.01 = 3

3.5 DIVISION

Division on the slide rule involves a logarithmic process by which the logarithm of the denominator is subtracted from the logarithm of the numerator, and the number corresponding to the differences between the logarithms is found on the D scale. Employing the example, 4/2, or 4 divided by 2, we set the denominator 2 on the C scale directly over the numerator 4 on the D scale and moving the hairline to the 1 on the C scale find the answer 2 below on the D scale. This is the exact reverse of multiplication (2 × 2 = 4). Also in division we subtract the log of 2 = 0.301 from the log of 4 = 0.602 to obtain the log

$$0.602 - 0.301 = 0.301$$

which corresponds to the number 2, which is the answer.

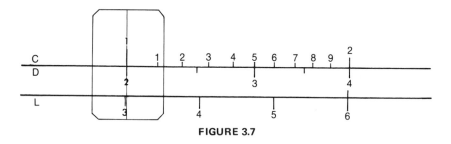

FIGURE 3.7

Another example will illustrate that the manipulation for division is the reverse of that for multiplication. In the problem 30/2 = 15 we also see that 15 × 2 = 30. On the slide rule to divide 30/2 we would set 2 on the C scale over 30 on the D scale and read 15 on the D scale under the 1 on the C scale, or to reverse the process and multiply 15 x 2, we would set the 1 on the C scale over 15 and move the hairline to 2 on the C scale to find the answer 30 below on the D scale.

Example

	Approximate answer	Correct answer
$\dfrac{52.5}{2.5} =$	$\dfrac{50}{2} = 25$	21.0

Note: The hairline should be used to line up properly the 52.5 on the D scale with the 2.5 on the C scale.

EXERCISES

Approximate answer

A. $\dfrac{111.1}{12.6} = \quad \dfrac{100}{10} = 10$

THE SLIDE RULE

B. $\dfrac{5280}{2.25} = \dfrac{5000}{2} = 2500$

C. $\dfrac{1.875}{0.0085} = \dfrac{2}{0.01} = 200$

3.6 COMBINATION — MULTIPLICATION AND DIVISION PROBLEMS

Frequently a calculation in the laboratory will involve both multiplication and division. These problems can be rapidly solved by the use of the slide rule. An example might be $\dfrac{4 \times 30}{6}$ where you could first carry out the multiplication and then divide the answer by 6 as follows: Set the 1 on the right of the C scale over 4 on the D scale; then move the hairline to the 3 on the C scale and find 120 below on the D scale. Now move the 6 on the C scale to the hairline over the 120 on the D scale, and the answer 20 is found on the D scale under the 1 on the right side of the C scale. To carry the process further, divide 4 × 30 × 8 by 6 × 3, or solve the problem $\dfrac{4 \times 30 \times 8}{6 \times 3}$. First multiply all the numbers by setting the 1 on the right side of the C scale over 8 on the D scale, then move the hairline to 4 on the C scale, which is over 32 on the D scale; next move the 1 on the left of the C scale over the hairline (32) and move the hairline to 3 on the C scale, which is over 960 on the D scale. Then to divide move the 6 on the C scale over the hairline (960) on the D scale and find 160 on the D scale under the 1 on the left of the C scale; finally move the 3 on the C scale over the hairline at 160 on the D scale, and by moving to the 1 on the right of the C scale you will find the answer 53.3 on the D scale.

Since so many problems will involve calculations such as $\dfrac{5 \times 22 \times 7}{3 \times 4}$, we will illustrate another slide rule method that may involve fewer manipulations; you may choose the method most easy to remember and to carry out for a particular problem. In this method you first divide and then multiply and alternate these operations until the problem is complete. First divide 5 by 3 by setting 3 on the C scale over 5 on the D scale; then multiply the answer on the D scale (167) by 22 merely by moving the hairline to 22 on the C scale, since the 1 on the C scale is already over the 167. Then divide the answer 367 on the D scale by 4 by setting 4 on the C scale over the hairline on 367 on the D scale; finally multiply the value 917 on the D scale by 7 by moving the hairline to 7 on the C scale, since again the 1 on the right side of the C scale is already set over the 917 on the D scale. The complete answer is found under the 7 on the C scale and is 641 on the D scale. The approximate answer is

$$\frac{5 \times 20 \times 7}{12} = \frac{700}{12} = 60$$

so the correct decimal setting is 64.1. With practice the method just described will save considerable time in a series of calculations.

Since these combination problems are commonly encountered in laboratory methods, we will work our way through another example before the practice problems. An extra long example would be:

$$\frac{4.25 \times 6.1 \times 23.3 \times 19.6}{0.38 \times 11.1 \times 8.9}$$

which we will work by **the alternate division-multiplication method.** First, divide 4.25 by 0.38 by setting 38 on the C scale over 425 on the D scale; then multiply the answer 1119 by 6.1 by moving the hairline to 61, since the 1 on the C scale is already set over 1119. Next divide the answer 682 by 11.1 by setting 11.1 on the C scale over the 682 under the hairline on the D scale; then multiply the answer 614 by 23.3 by setting the 1 on the right side of the C scale over 614 and moving the hairline to the 23.3 on the C scale. The answer 143 is next divided by 8.9 by moving the 8.9 on the C scale over the hairline at 143; the answer 1609 is finally multiplied by 19.6 by moving the 1 (on the left of the C scale) over the 1609 and moving the hairline to the 19.6 on the C scale to give the final answer 315. The approximate answer is $\frac{4 \times 6 \times 20 \times 20}{0.4 \times 10 \times 10} = 240$, so the correct decimal setting is 315.

EXERCISES

Approximate answer

A. $\dfrac{35 \times 3}{9}$ $= \dfrac{30 \times 3}{10} = 9$

B. $\dfrac{5 \times 12.5 \times 55}{4 \times 11}$ $= \dfrac{5 \times 10 \times 60}{5 \times 10} = 60$

C. $\dfrac{3.14 \times 67.2}{5.25 \times 8.95}$ $= \dfrac{3 \times 70}{5 \times 10} = 4.2$

THE SLIDE RULE

D. $\dfrac{510 \times 6.1 \times 21.3}{4.98 \times 0.071} = \dfrac{500 \times 6 \times 20}{5 \times 0.01} = 120{,}000$

E. $\dfrac{17.2 \times 22.2 \times 9.3}{898 \times 0.212} = \dfrac{20 \times 20 \times 10}{1000 \times 0.2} = 20$

3.7 SQUARES AND SQUARE ROOTS

Both the **squares** and **square roots** of numbers are commonly encountered in quality control work in the laboratory. The manipulations to solve this type of problem on the slide rule **involve only the D and A scales** without the use of the sliding C scale.

To find the square of a number set the hairline over that number on the D scale and read the answer on the A scale under the hairline. You will observe that the A scale is a double scale, containing the numbers on the C or D scale on each half. The squaring of simple numbers such as 2 and 4 will illustrate the utility of the double scale. Set the hairline over the 2 on the D scale and read 4 under the hairline on the A scale ($2^2 = 4$). Also set the hairline over the 4 on the D scale and read 16 under the hairline on the A scale ($4^2 = 16$). To square 29.2 set the hairline on 292 on the D scale and read 852 on the A scale. To set the decimal point round off 29.2 to 30, square it mentally and obtain 900. The answer therefore, is 852.

Examples

Square the following numbers

Approximate answer

A. 3.67 = 3 × 3 = 9

B. 22.2 = 20 × 20 = 400

C. 6.98 = 7 × 7 = 49

D. 816 = 800 × 800 = 640,000

To find the square root of a number set the hairline over that number on the A scale and read the answer under the hairline on the D scale. Even though square roots are found by a method opposite to that used for finding the square of a number, the double A scale poses some problems. To decide which half of the A scale to use the simple scheme outlined below may be followed.

1. If the number is 1 or greater, space off groups of two digits from the decimal point to your **left**. If the last group to the left has one digit, use the left half of A; if the last group to the left has two digits, use the right half of A. For example, to find the square root of 676.0 divide the number two digits to the left of the decimal point (6/76); you will find one digit in the last group to the left. Setting 676 on the left half of A, the answer is 26. It is obvious that $26^2 = 676$ so the answer is correctly set at 26. In a quality control problem the sum of the squares may be 1459.0. To find the square root of 1459.0 first divide the digits to the left of the decimal point in groups of two (14/59). Since the last group to the left has two digits use the right side of the A scale. Setting 1459 on the right side of the A scale the answer under the hairline on the D scale is 382. To decide on the decimal point round off 382 to 400 and try $400 \times 400 = 16,000$, which is incorrect; so try $40 \times 40 = 1600$, which sets the correct number as 38.2.

2. If the number is less than 1, space off groups of two digits from the decimal point to your **right**. If the first group with a significant figure to right of the decimal point contains one number, use the left half of the A scale. If the first group contains two numbers, use the right half of the scale. To find the square root, for example, of 0.0096 again space off in groups of two (0.00/96). Since there are two numbers in the group with significant figures, we use the right half of scale A, setting 96 and reading 98 on the D scale.

It is more difficult to set the decimal point in square roots of numbers less than one so an arbitrary scheme is used. Starting at the decimal point of 0.0096 place the first digit of the answer under the first group that contains a significant number, then the second digit of the answer is placed under the next group to the right and so on. A zero is placed under any group of two zeros as follows: 0.00/96/ . The answer would then be 0.098. A quick check could be
   0 9 8

made by multiplying $0.09 \times 0.09 = 0.0081$, proving the answer in the correct range of decimal point. To try a very small number find the square root of 0.000004120. First divide into groups 0.00/00/04/12, and since there is one number in the first group with a significant figure, set 412 on the left side of the A scale, reading 2032 on the D scale. Set up the original figure as 0.00/00/04/12/00/00 with the answer below the groups. The correct square root
   0 0 2 0 3 2

is then 0.002032; to check the value $0.002 \times 0.002 = 0.000004$.

EXERCISES

                                                             Trial Check

E. The square root of 66 = 812 $8 \times 8 = 64$

F. The square root of 567 = 238 $20 \times 20 = 400$

THE SLIDE RULE   **31**

G. The square root of 0.0435 = 2085   0.2 × 0.2 = 0.04

H. The square root of 6.5 = 255   2 × 2 = 4

I. The square root of 0.000168 = 1298   0.01 × 0.01 = 0.0001

J. The square root of 5120 = 717   70 × 70 = 4900

3.8 CONVERSION AND PROPORTION

In the laboratory we use almost exclusively the metric system for units of length, weight and volume. In the home and supermarket units of measurement differ from those used in the laboratory, and we are often called upon to convert one unit to another. For example, the relation between a few common units of measurement are shown in the following list.

        1 inch = 2.54 centimeters (1 cm = 0.3937 in.)
        1 yard = 0.9144 meters (1 m = 1.094 yd.)
        1 mile = 1.609 kilometers (1 km = 0.6214 mi.)
        1 fluid ounce = 29.57 milliliters (1 ml = 0.0338 oz.)
        1 gallon = 3.785 liters (1 l = 0.2641 gal.)
        1 grain = 64.8 milligrams (1 mg = 0.01543 gr.)
        1 pound = 453.6 grams (1 gm = 0.002205 lb.)

To convert inches to centimeters set the 1 on either end of the C scale over the 2.54 on the D scale. Then set the hairline over the required number of inches and read the answer in centimeters on the D scale. For example, 4 in. = _____ cm? Set the 1 on the right side of the C scale over 2.54 on the D scale, move the hairline to 4 on the C scale and read 10.16 cm on the D scale. The speedometer on an American automobile read 60 mi/hr on a British highway. What was the automobile's rate of speed in km/hr? Set the 1 on the left side of the C scale over the 1.609 on the D scale, move the hairline to 60 on the C scale and read 96.6 km/hr on the D scale. A 1/2 pound bottle of NaCl contains how many grams of NaCl? Set the 1 on the right side of the C scale over 453.6 on the D scale, move the hairline to 5 (0.5) on the C scale and read 226.8 gm on the D scale. In general, when you set the 1 on either end of the C scale over a conversion factor, all the values on the C scale are in the units represented by the conversion factor, and all the values on the D scale are in the units to which you are converting.

EXERCISES

Convert:

A. 23.5 in to cm

B. 5 yds to m

C. 0.3 oz to ml

D. 5 gr to mg

E. 120 lbs to kg

Proportions are similar to conversions, but in proportions set conversion factors are usually not given. A statement may be made that 5 yds are equivalent to 4.57 m; the question may then be asked, what is the equivalent in meters to 1.5 yds? A proportion could be set up:

$$5:4.57 = 1.5:X \text{ or } \frac{5}{4.57} = \frac{1.5}{X}$$

Solving for X

$$X = \frac{4.57 \times 1.5}{5}$$

The value X in a proportion can always be solved for by multiplication and division, as in a regular problem, or the slide rule can be used in a simplified scheme. The proportion may be set up on the C and D scales as

$$C:D = C:D \text{ or as } \frac{C}{D} = \frac{C}{D}$$

For the example 5:4.57 = 1.5:X, set 5 on the C scale over 4.57 on the D scale and move the hairline to 1.5 on the C scale, reading the answer 1.368 yds on the D scale. Or a problem may be stated that 3 in equal 76.2 mm; how many inches are there in 300 mm? This proportion would be set up as:

$$3:76.2 = X:300 \text{ or } \frac{3}{76.2} = \frac{X}{300}$$

Again C:D = C:D or C/D = C/D; therefore, set 3 on the C scale over 76.2 on the D scale and move the hairline to 300 on the D scale, reading the answer 11.82

THE SLIDE RULE   **33**

in, on the C scale. If the second value on the D scale does not occur under the sliding C scale and thus cannot be read directly as in the proportion 3:76.2 = X:200, again set 3 on the C scale over the 76.2 on the D scale and observe that 200 is not under the sliding C scale. To complete the proportion set the hairline on the 1 on the left of the C scale over 254 on the D scale, move the 1 on the right of the C scale to the hairline, then move the hairline to 200 on the D scale and read the answer 7.88 in on the C scale.

You may have observed that in every proportion we are merely setting up the slide rule directly as the proportion stated. From the proportion C/D = C/D we transfer to the two scales of the slide rule:

$$\frac{C}{D} \quad \frac{11.82\ (X)}{300} \quad \frac{3}{76.2} \quad \frac{C}{D}$$

OR

$$\frac{C}{D} \quad \frac{7.88\ (X)}{200} \quad \frac{3}{76.2} \quad \frac{C}{D}$$

OR

$$\frac{C}{D} \quad \frac{3 \quad 35.45\ (X)}{76.2 \quad 900} \quad \frac{C}{D}$$

EXERCISES

F. If 10 lbs = 4.54 kg, how many lbs = 50 kg?

G. If 5 mi = 8.045 km, 8 mi = how many km?

H. 9:20 = 5:X

I. 14:56 = X:3

J. 90:75 = X:25

3.9 THE CIRCULAR SLIDE RULE

A circular slide rule of reasonable size — for example the Dietzen Binary Slide Rule, which is 8 inches in diameter — has some distinct advantages over the common 10 inch long slide rule. The most **important advantage** is **the extra length of the commonly used scales.** The C scale used in the majority of calculations is just over 25 inches long and the L, or log scale, is almost 19 inches long. This facilitates setting of the hairline indicators and increases the accuracy of the calculations. Some idea of the advantage of a 25 inch C scale can be obtained from a study of the subdivisions of the major numbers on the scale. The largest space between 1 and 2 is not only divided into tenths, but each tenth is divided into 20 divisions, or a total of 200 divisions, with each individual division larger than one of the 100 divisions between 1 and 2 on the 10 inch long slide rule. Numbers from 2 to 5 are divided into tenths with each tenth further divided into tenths, while numbers from 5 to 10 are divided into tenths with each tenth further divided into fifths. In addition all the 100 tenths divisions between 1 and 10 are identified with a printed number corresponding to the particular tenth between any two numbers.

The front side of the slide rule has the C scale on the periphery or outer scale; under it is the CI or inverted C scale. Then comes the A scale for squares and square roots, next the K scale for cubes and cube roots, and the fifth scale in from the periphery is the L or log scale. There are two plastic hairline indicators on the front side of the slide rule. The **long indicator** is referred to as **L**, while the **short indicator** is called **S**. It will be noted that **whenever S is moved L remains stationary**, but that **when L is moved, S moves with it.** Whenever L is moved in solving a problem be sure that nothing interferes with the free movement of S. **L always gives the answer to the problem.**

Multiplication. To multiply 1.5×2, first set L at 1.5 and S at 10. Turn L until S is at 2 and L will indicate the answer, 3. The new positions of L and S after moving S to 2 are indicated as L' and S' in the diagram. Practice multiplication by working the following problems.

EXERCISES

A. $185 \times 4.46 =$ _____

B. $92 \times 6.85 \ =$ _____

To multiply more than two numbers you merely repeat the simple process outlined above, as in the problem, $3 \times 6 \times 7$. Set L at 3 and S at 10. Turn L counterclockwise until S is at 6 (answer 18). Then move S back to 10 and once again move L counterclockwise until S is at 7; read the answer 126 under the L indicator. Practice this technique with the following problems.

EXERCISES

C. $7.2 \times 0.923 \times 11.4 = 8 \times 1 \times 10 = 80$

THE SLIDE RULE

D. $52.1 \times 18.9 \times 5.7 \times 9.3 = 50 \times 20 \times 5 \times 10 = 50,000$

Division. To divide 3/2, first set L at 3 and S at 2. These positions are indicated as L' and S' in the diagram. Turn L' until S' is at 10 and L will indicate the answer, 1.5. Practice division by working the following problems.

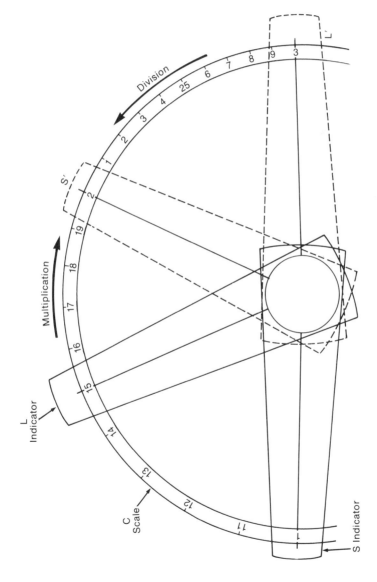

FIGURE 3.8

EXERCISES

E. $\dfrac{111.6}{9.7} = \dfrac{100}{10} = 10$

F. $\dfrac{5280}{3.141} = \dfrac{5000}{3} = 1667$

Combination Multiplication and Division Problems. The simplest way to solve combination type problems on a circular slide rule is to carry out **alternate, divisions and multiplications.** As an example, solve the problem $\dfrac{4 \times 30 \times 8}{6 \times 5}$. First divide 4 by 6 by setting L on 4 and S on 6 and turning L until S is at 10; then multiply the answer 0.667 by 30 by turning L until S is at 30. Next divide the answer 20 by 5 by setting S on 5 and turning until S is at 10. Finally multiply the answer 4 by 8 by turning L until S is at 8, giving the final answer 32.

After you are familiar with the circular slide rule you will discover an important shortcut in these combination problems. Let us try the short version; divide 4 by 6 by setting L on 4 and S on 6; then instead of returning S to 10, multiply by 30 by turning L until S is at 30. Next divide the answer 20 by 5 by setting S on 5; then again instead of returning S to 10, multiply by 8 by turning L until S is at 8, giving the final answer 32 at L. An extra long problem such as $\dfrac{4.25 \times 6.3 \times 22.8 \times 19.7}{0.52 \times 11.3 \times 8.8}$ could be solved by the shorter method as follows: First divide 4.25 by 0.52 by setting L on 4.25 and S on 0.52; then multiply by 6.3 by turning L until S is at 6.3. Next divide the answer (51.5) on L by 11.3 by setting S on 11.3; then multiply by 22.8 by turning L until S is at 22.8. Finally divide the answer (104.0) on L by 8.8 by setting S on 8.8, then multiply by 19.7 by turning L until S is at 19.7, giving the answer 232.8. Practice the short method by solving the following problems.

EXERCISES

G. $\dfrac{33 \times 5 \times 3}{9 \times 2} = \dfrac{30 \times 5 \times 3}{10 \times 2} = 22.5$

H. $\dfrac{68.3 \times 3.141 \times 11.2}{5.67 \times 9.12} = \dfrac{70 \times 3 \times 10}{5 \times 10} = 42$

THE SLIDE RULE

Square and Square Roots. The technique of finding the square or square root of a number is exactly the same as on a 10 inch slide rule. The C and A scales (first and third) are used on the circular slide rule. Set L on any number on the C scale and read the square under the hairline on the A scale. For example, square 6.89 = 7 X 7 = 49, correct answer 47.5. In solving square root problems, if the answer is 1 or greater, space off groups of two digits from the decimal point to your left. If the last group to the left has one digit, use the first half of the A scale; if the last group to the left has two digits, use the last half of the A scale. As an example, to find the square root of 450.0 space off two digits to the left and find one digit in the last group (4/50.0). Setting 450.0 on the first half of A, find 212 on the C scale. Since 20 X 20 = 400 the correct answer is 21.2. If the number is less than 1, space off groups of two digits from the decimal point to your right. If the first group with a significant figure to the right of the decimal point contains one number, use the first half of the A scale; if the group contains two numbers, use the second half of the A scale. The square root of 0.000860 would be obtained by dividing into groups as 0.00/08/60, and since there is one number in the first group with a significant figure, set 860 on the first half of the A scale, reading 293 on the C scale. Set up the original figure as 0.00/08/60/00 with the answer below the groups, as with the 10 inch slide rule.

   0 2 9 3

The correct square root is then 0.0293; to check the value, 0.03 X 0.03 = 0.0009.

EXERCISES

                                                    Trial check

I. The square root of 73 = 854   8 X 8 = 64

J. The square root of 0.00445 = 667   0.07 X 0.07 = 0.0049

Conversion and Proportion. The main difference between conversion and proportion problems on the circular slide rule is the technique of setting the scales and reading the answers to the problems. A few examples utilizing the circular slide rule will suffice to make the point. To convert 6 inches to centimeters with the relationship 1 in = 2.54 cm set L on 2.54 and S on 10; then move L until S is at 6, and L will indicate the answer, 15.24 cm. A 5 pound bottle of glucose contains how many grams of glucose? Since 1 lb = 453.6 gm, set L on 453.6 and S on 10; then move L until S is at 5, and L will indicate the answer, 2268 gm. Proportions are simple to calculate once the proportion is correctly set up. If 8 yds is equivalent to 7.32 m, how many meters are equivalent to 2.5 yds? The proportion could be set up as:

$$8:7.32 = 2.5:X \text{ or } \frac{8}{7.32} = \frac{2.5}{X} \text{ or } X = \frac{7.32 \times 2.5}{8}$$

A simple division followed by a multiplication would be solved as follows: Divide 7.32 by 8 by setting L on 7.32 and S on 8; then multiply by 2.5 by turning L until S is at 2.5, reading the answer, 2.287, meters on L.

EXERCISES

K. If 20 lbs = 9.09 kg, how many lbs = 70 kg?

L. 95:80 = X:32

M. If 60 mi = 96.54 km, 150 mi = how many kilometers?

ADDITIONAL EXERCISES

1. Carry out the following calculations with a slide rule.

    a. 220×13

    b. 3.3×69

    c. $96.2 \times 4.5 \times 21.3$

    d. $22.2 \times 316 \times 42.4$

    e. $6.1 \times 11.5 \times 206 \times 96.8$

2. Work the following problems with a slide rule.

    a. $\dfrac{5280}{3}$

    b. $\dfrac{5 \times 21 \times 9}{7 \times 2}$

c. $\dfrac{17.5 \times 110.1 \times 68}{51.6 \times 22.2}$

d. $\dfrac{0.198 \times 98.8 \times 12.3}{0.202 \times 53.1}$

e. $\dfrac{3.14 \times 5280 \times 1.5 \times 9.9}{23.1 \times 2.2 \times 8.5}$

3. Calculate the following with a slide rule.

   a. 31.3^2

   b. 1.51^2

   c. $\sqrt{552}$

   d. $\sqrt{0.056}$

   e. $\sqrt{67.5}$

4. Convert the following units.

   a. 141 in to cm

   b. 36 in to cm

   c. 200 lbs to kg

   d. 130 lbs to kg

   e. 10 gr to mg

   f. 1.25 gr to mg

   g. 60 mi to km

   h. 8 oz to ml

5. Calculate the following proportions:

   a. 80:20 = X:10

   b. If 50 lbs = 22.7 kg, how many pounds = 50 kg?

c. If 5 gr = 324 mg, how many grains = 500 mg?

d. If 176.0 yds = 0.1 mi, how many yards = 5 mi?

e. If 2 oz = 59.14 ml, how many ounces = 1000 ml?

CHAPTER 4

CONCENTRATION OF SOLUTIONS

In the preceding sections we have been exposed to, and in some instances reviewed, the expression of numbers, powers of ten, roots, logarithms and the use of the slide rule. If you are willing to thoughtfully practice the examples that were given to enable you to **become proficient in arithmetical common sense and in the use of logarithms and the slide rule, you will possess the tools to solve any of the problems in laboratory mathematics.**

There are many occasions in the laboratory when it is necessary for the technician to prepare solutions to be used in routine or research methods and reactions. Obviously it is important that every solution be correctly prepared, and to complete such an assignment you must first understand the expressions of concentration and then be able to correctly calculate the quantities of solvent and solute required to make the solution.

4.1 PERCENTAGE SOLUTIONS

Per cent means parts per hundred and is commonly used by banks as per cent interest. If the banks pay 5 per cent interest per year on savings, a deposit of $100 would earn $5 interest in one year. In the laboratory per cent is expressed as grams per hundred or milliliters per hundred. The three most common percentage solutions used in the laboratory are:
1. grams of solute in 100 milliliters of solution (W/V)
2. grams of solute in 100 grams of solution (W/W)
3. milliliters of liquid in 100 milliliters of solution (V/V)

W/V signifies weight per volume, W/W weight per weight and V/V volume per volume. In most laboratories the **weight per volume solutions** are used because the solute is important and the solvent is merely used as the vehicle for the solute. The physical chemist prefers to weigh both the solute and solvent to prepare weight per weight solutions because the concentration does not change with changes in temperature. The volume per volume solutions are the least accurate and are used when the solute is a liquid, as in alcohol solutions.

The preparation of a 5 per cent solution of sodium chloride, for example, on a weight per volume basis would consist of dissolving 5 gm of sodium chloride in water and then diluting the solution to a final volume of 100 ml. On a weight per weight basis, 5 gm of sodium chloride would be added to 95 gm of water and dissolved to form 100 gm of solution. A solution of two liquids on a volume per volume basis, as, for example, a 10 per cent solution of alcohol, would be prepared by adding 10 ml of alcohol to water and diluting to a final volume of 100 ml. In simple calculations involving percentage solutions first determine the quantity of solute contained in 100 ml of solution. It then becomes relatively easy to calculate the quantity of solute present in a given volume.

Examples

Find the weight of glucose that must be used to prepare 800 ml of a 5 per cent solution. First we can state that a 5 per cent solution contains 5 gm of glucose in 100 ml of solvent. Since 800 ml is required 800/100 = 8 times the amount in 100 ml or 8 × 5 = 40 gm of glucose that must be used to prepare 800 ml of a 5 per cent solution.

What volume of isotonic saline (0.85% NaCl) can be prepared from 42.5 gm of NaCl? In outline form:

1. An 0.85 per cent solution contains 0.85 gm in 100 ml.

2. Since 42.5 gm are to be used, $\dfrac{42.5}{0.85}$ = 50 times the amount in 100 ml.

3. 50 × 100 ml = 5000 ml of 0.85 per cent NaCl solution contains the 42.5 gm.

EXERCISES

A. Find the weight of sodium sulfate that must be used to prepare 250 ml of a 23 per cent solution.

CONCENTRATION OF SOLUTIONS

B. A weighed quantity of NaCl (50.0 gm) was accidently transferred to a beaker containing approximately 200 ml of distilled water. How would you proceed to prepare a 5 per cent NaCl solution from the contents of the beaker?

a. A 5 per cent solution contains 5 gm in 100 ml.

b. Since 50 gm NaCl are in the beaker, $\frac{50}{5} = 10$ times the amount in 100 ml.

c. 10 × 100 ml = 1000 ml of 5 per cent NaCl solution equivalent to 50 gm.

d. Quantitatively transfer the NaCl solution from the beaker to a liter volumetric flask and dilute to the 1000 ml mark with water.

Before proceeding to problems that involve molar or normal solutions and even before discussing more complicated percentage solutions we must consider a few accessory points such as specific gravity of solutions and inorganic salts having varying numbers of molecules of water of hydration.

4.2 SPECIFIC GRAVITY

The density of both solids and liquids is usually expressed in grams per milliliter (gm/ml). The density of mercury, for example, is 13.6 gm/ml, meaning of course that each milliliter of mercury weighs 13.6 grams. The **specific gravity** of a solid or a liquid is a number that expresses the ratio of the weight of the solid or liquid to the weight of an equal volume of water at 4°C taken as a standard.

$$\text{Specific gravity} = \frac{\text{Weight of solid or liquid}}{\text{Weight of an equal volume of water at } 4°C}$$

The density of water changes very little between 4° and 30°C; therefore, the value 1.00 gm/ml may be used as a standard for specific gravity measurements at ordinary laboratory temperatures. The specific gravity of a solid or liquid is the same when any system of units is used, because it expresses the ratio of the weight of the solid or liquid to the weight of an equal volume of water. Specific gravity values are therefore expressed as a number without any units. For example, the specific gravity of normal urine varies from 1.008 to 1.030.

Examples

Find the density and specific gravity of ethyl alcohol if 8.0 ml weights 6.40 gm.

1. If 8.0 ml weighs 6.4 gm, then 1 ml weighs $\frac{6.4}{8.0} = 0.80$ gm.

2. The density is therefore 0.80 gm/ml and the specific gravity is
$$\frac{0.80 \text{ gm/ml}}{1.00 \text{ gm/ml } (H_2O)} = 0.80$$

A volume of urine (500 ml) weighs 510.0 gm. What is the specific gravity of the urine specimen?

Since 500 ml of urine weighs 510 gm and 500 ml of water weighs 500 gm, $\frac{510}{500} = 1.020$, the specific gravity of the urine.

EXERCISES

A. What would be the volume of 20 kg of a glycerol solution that had a specific gravity of 1.55?

B. A small volume, 25 ml, of a heavy liquid weighs 340 gm. What is the specific gravity of the liquid?

C. Concentrated hydrochloric acid contains 37.5 per cent HCl by weight and has a specific gravity of 1.19. Calculate the weight of HCl in 10.0 ml of concentrated hydrochloric acid.

4.3 HYDRATES AND WATER OF HYDRATION

In addition to anhydrous salts you will often find reagent bottles of hydrated solid inorganic salts, as, for example, $CuSO_4 \cdot 5\ H_2O$, $BaCl_2 \cdot 2\ H_2O$, and so on. When preparing solutions we usually weigh quantities of the solid solute to be dissolved in a definite volume of final solution. To prepare a 5 per cent solution of $CuSO_4$ when only the hydrate $CuSO_4 \cdot 5\ H_2O$ is available, we must calculate the weight of the hydrate that contains 5 gm of $CuSO_4$

CONCENTRATION OF SOLUTIONS 45

(anhydrous). This is accomplished by dividing the formula weight of $CuSO_4 \cdot 5 H_2O$ by the formula weight of $CuSO_4$ and multiplying by 5, as follows:

$$\frac{CuSO_4 \cdot 5 H_2O}{CuSO_4} = \frac{63.5 + 32 + 64 + 10 + 80}{63.5 + 32 + 64} = \frac{249.5}{159.5} \times 5 = 7.83 \text{ gm}$$

It would be necessary to weigh 7.83 gm of $CuSO_4 \cdot 5 H_2O$ to obtain 5 gm of the anhydrous salt to dissolve in water for a total volume of 100 ml of a 5 per cent solution.

Example

One pound (453.6 gm) of $BaCl_2 \cdot 2 H_2O$ is equivalent to what weight of $BaCl_2$?

1. 1 gm of $BaCl_2 \cdot 2 H_2O$ is equivalent to:

$$\frac{BaCl_2}{BaCl_2 \cdot 2 H_2O} = \frac{137.3 + 71}{137.3 + 71 + 36} = \frac{208.3}{244.3} = 0.853 \text{ gm}.$$

2. $453.6 \times 0.853 = 386$ gm.

3. One pound of $BaCl_2 \cdot 2 H_2O$ contains 386 gm of $BaCl_2$.

EXERCISES

A. A pound of washing soda, $Na_2CO_3 \cdot 10 H_2O$, contains what percentage of water?

B. Find the weight of $Na_2SO_4 \cdot 7 H_2O$ that must be used to prepare 250 ml of a 23 per cent solution of Na_2SO_4.

4.4 MOLAR SOLUTIONS

A gram molecular weight of a substance is known as a **mole. A solution that contains 1 mole of the solute in 1 liter is called a molar solution.** The abbreviation for molar is M; thus, a one molar solution can be expressed as 1 M.

Example

What weight of H_2SO_4 is required to prepare 400 ml of a 3 M solution?

1. The molecular weight of H_2SO_4 is 98; therefore, one liter of 3 M solution contains $3 \times 98 = 294$ gm H_2SO_4.

2. If one liter contains 294 gm, 400 ml contains $0.4 \times 294 = 117.6$ gm of H_2SO_4.

When working with molar solutions if you first determine the quantity of solution in a liter, it is then easy to calculate fractions or multiples of that value. In general chemistry we learned that several salts and hydroxides do not have true molecular weights. This fact makes it difficult to express all compounds in moles, although a formula may be written for any given compound. Some chemists prefer formal solutions to molar solutions and formality instead of molarity. A formal solution, abbreviated F, contains a gram formula weight in a liter of solution and in many instances is the same as a molar solution. Since it is common practice in most laboratories to express concentration in terms of molarity, we will retain the molar designation.

In the modern laboratory the use of micromethods, the study of subcellular components and the vast scope of enzyme methodology have given rise to the expressions of concentrations in **millimoles**, mM, and **micromoles**, μM. A millimole is the molecular weight in milligrams instead of grams, and micromole is a microgram molecular weight. These small concentrations are also often expressed as 1×10^{-3} M, which equals 1 mM, or 1×10^{-6} M, which equals 1 μM. Another useful relation is the fact that **each milliliter of a 1 M solution contains 1 mM of the solute.**

Example

How many grams of HCl would be required to prepare 500 ml of a 1×10^{-3} M solution?

1. A 1×10^{-3} M solution = 1 mM HCl solution, which contains a milligram molecular weight per liter or 36.5 mg/l.

2. If one liter contains 36.5 mg, 500 ml contains 18.25 mg, and 0.01825 gm of HCl is needed to prepare 500 ml of a 1×10^{-3} M solution.

CONCENTRATION OF SOLUTIONS 47

EXERCISES

A. What weight of NaOH is required to prepare 3000 ml of a 0.5 M solution?

B. How many micrograms of NaCl would 300 ml of a 1×10^{-6} M solution contain?

C. A solution contains 3.65 gm of HCl in a liter. How many millimoles of HCl does it contain?

4.5 MOLAL SOLUTION

A molal solution contains 1 mole of the solute in 1000 gm of solvent. In comparison, a molar solution has a final volume of 1000 ml, whereas the volume of a molal solution exceeds 1000 ml. Physical chemists often use molal solutions because temperature changes do not affect the concentration as they do in molar solutions. Most laboratories do not weigh the solvent and also do not require the exacting properties of molal solutions in their reactions. We shall not, therefore, give examples of the preparation or calculation of molal solutions.

4.6 NORMAL SOLUTION

A normal solution contains one gram-equivalent weight of the solute in 1 liter of solution. A gram-equivalent weight is the quantity of a substance that will replace or react with 1.008 gm of hydrogen. The gram-equivalent weight of an acid would be the weight equivalent to 1.008 gm of replaceable hydrogen. An equivalent weight of a base is the weight that combines with 1.008 gm of replaceable hydrogen. For example, 1 mole of HCl, 1/2 mole of H_2SO_4, 1 mole of NaOH, 1/3 mole of $Al(OH)_3$ and 1/2 mole of K_2SO_4, each in 1 liter of solution would represent normal solutions of these compounds. A one normal solution is abbreviated 1 N solution. Since a milligram-equivalent weight of a substance would be a milliequivalent, a normal solution of a compound would contain 1 milliequivalent of solute in 1 ml of solution.

Examples

How many milliliters of concentrated HCl (specific gravity 1.19 and 37.50 per cent HCl by weight) would be required to prepare 15 l of 0.1 N acid?

1. The equivalent weight of $HCl = \frac{36.5}{1} = 36.5$; therefore, 1 liter of 0.1 N solution contains $\frac{36.5}{10} = 3.65$ gm HCl and 15 l would contain $15 \times 3.65 = 54.75$ gm.

2. One milliliter of concentrated acid contains $0.375 \times 1.19 = 0.446$ gm HCl. Therefore, 15 liters of 0.1 N HCl requires $\frac{54.75}{0.446} = 122.8$ ml of concentrated acid.

How many milliequivalents of a solute are contained in 500 ml of a 3 N solution?

One liter (1000 ml) of a 1 N solution contains 1000 milliequivalents; therefore, 1000 ml of a 3 N solution contains 3×1000 or 3000 milliequivalents and 500 ml contains $\frac{3000}{2} = 1500$ milliequivalents in 500 ml of a 3 N solution.

EXERCISES

A. How many milliliters of concentrated sulfuric acid (specific gravity 1.84 and 98 per cent H_2SO_4 by weight) are required to prepare 3 l of 0.2 N solution?

B. What is the normality of isotonic saline (0.85 per cent NaCl)?

C. Given 500 ml of a 40 per cent solution of NaOH, to what volume of 1 N NaOH is this equivalent?

4.7 DILUTION OF SOLUTIONS

All of the solutions that we have previously considered are known as volumetric solutions and all contain a definite amount of solute in a fixed volume of solution. In percentage, molar, molal or normal solutions, the amount of solute contained in a given volume of solution is equal to the product of the volume times the concentration. When a solution is diluted, its volume is increased and its concentration is decreased; however, the total amount of solute remains unchanged. Two solutions with different concentrations that contain the same amount of solute can therefore be related to each other as follows:

CONCENTRATION OF SOLUTIONS

$$\text{Amount of solute}_1 = \text{Volume}_1 \times \text{Concentration}_1$$
$$\text{Amount of solute}_2 = \text{Volume}_2 \times \text{Concentration}_2$$

Since the amount of solute$_1$ = amount of solute$_2$, then:

$$\textbf{Volume}_1 \times \textbf{Concentration}_1 = \textbf{Volume}_2 \times \textbf{Concentration}_2$$

If the volume and concentration on both sides of the equation are expressed in the same units, this relationship will apply in many calculations. For example:

$$\text{ml} \times \text{N} = \text{ml} \times \text{N}$$

$$l \times M = l \times M$$

$$\text{ml} \times \mu M = \text{ml} \times \mu M$$

Examples

How much water should be added to 2 l of 0.12 N hydrochloric acid to prepare a solution of 0.10 N HCl?

From the equation:
$$\text{ml}_1 \times \text{N}_1 = \text{ml}_2 \times \text{N}_2$$
Let X = volume of 0.10 N solution (ml$_1$)
$$X_{ml} \times 0.10 \text{ N} = 2000 \text{ ml} \times 0.12 \text{ N}$$
or $X = \dfrac{2000 \times 0.12}{0.10} = 2400$ ml of 0.10 N HCl and 2400 − 2000 (the 2000 ml of 0.12 N) = 400 ml of water that should be added to the 2 l of 0.12 N HCl to prepare 0.10 N HCl.

What volume of 10 per cent NaCl solution would be required to prepare 800 ml of a 0.85 per cent solution?

From the equation:
$$\text{ml}_1 \times \text{per cent}_1 = \text{ml}_2 \times \text{per cent}_2$$
Let X = volume of 10 per cent solution (ml$_1$)
$$X_{ml} \times 10 \text{ per cent} = 800 \text{ ml} \times 0.85 \text{ per cent}$$
or $X = \dfrac{800 \times 0.85}{10} = 68$ ml of 10 per cent NaCl required.

EXERCISES

A. Given 50 ml of a 12 μM solution of $BaCl_2$, what volume of 3 μM $BaCl_2$ could be prepared from this solution?

B. Fifteen liters of approximately 0.1 N NaOH were prepared from a stock solution. An accurate titration indicated a normality of 0.1055 N. How much water should be added to the 15 liters to prepare 0.100 N NaOH?

C. What volume of a stock solution of 20 per cent phosphoric acid would be required to prepare 600 ml of a 1.5 per cent solution?

In the clinical laboratory dilutions that are commonly encountered are the **dilution of a specimen** for accurate analysis and the **dilution of a standard** to correspond to the concentration in the serum. These dilutions are expressed as a ratio, such as 1:10, which refers to 1 unit of the original solution diluted to a final volume of 10 units. The diluted solution would be 1/10 the concentration of the original solution, and its concentration would be calculated by multiplying the concentration of the original solution by the dilution expressed as a fraction.

Examples

A 100 mg/100 ml glucose solution is diluted 1:5. What is the concentration of the dilute solution?

Original solution = 100 mg/100 ml × the dilution 1/5 = 20 mg/100 ml

A serum specimen was diluted 1:2, and this diluted solution was further diluted 1:4. If the concentration of the final diluted solution was 50 mg/100 ml, what was the concentration in the serum specimen? To calculate the original concentration we must reverse the dilution fractions:

$$50 \text{ mg}/100 \text{ ml} \times \frac{2}{1} \times \frac{4}{1} = 400 \text{ mg}/100 \text{ ml serum concentration}$$

To check: $400 \text{ mg}/100 \text{ ml} \times \frac{1}{2} \times \frac{1}{4} = 50 \text{ mg}/100 \text{ ml}$

Serum is often diluted 1:10 in the preparation of a protein-free filtrate. If 1 ml of the filtrate is used in a spectrophotometric method and 1 ml of a standard solution is carried through the same method, the standard solution must be diluted 1:10 if it is to represent the serum concentration. Under these conditions

CONCENTRATION OF SOLUTIONS

what standard concentration would correspond to a serum glucose level of 200 mg/100 ml? Serum diluted 1:10, so $200 \times \frac{1}{10} = 20$ mg/100 ml the concentration of the standard used in the method.

EXERCISES

D. A 500 mg/100 ml solution was diluted 1:10, and this diluted solution was further diluted 1:5. What is the concentration of the final solution?

E. A serum specimen was diluted 1:10, and this diluted solution was further diluted 1:3. If the concentration of the final diluted solution was 20 mg/100 ml, what was the concentration in the serum specimen?

4.8 CONVERSION PROBLEMS

Problems in the laboratory that involve **conversion from one unit of concentration to another** are often encountered. The preparation of normal or molar solutions from percentage solutions or the expression of mg/100 ml units as milliequivalents/liter are common conversion problems. All the principles that have been discussed under solutions will be applied to conversion problems. In general, the first step is to convert the given units to the desired units of concentration and then to calculate any dilution problem using similar units.

Examples

How many milliliters of 22 per cent Na_2SO_4 would be required to prepare 500 ml of a 0.5 M solution?

1. A 22 per cent Na_2SO_4 solution contains 22 gm/100 ml $\times$ 10 = 220 gm/l.

2. A 1 M Na_2SO_4 solution contains one gram molecular weight or 142 gm/l. Therefore, 22 per cent $Na_2SO_4 = \frac{220 \text{ gm/l}}{142 \text{ gm/l}} = 1.55$ M.

3. From $ml_1 \times M_1 = ml_2 \times M_2$

Let X = volume of the 1.55 M solution, therefore

$$X_{ml} \times 1.55 \text{ M} = 500 \text{ ml} \times 0.5 \text{ M}$$

or $X = \dfrac{500 \times 0.5}{1.55} = 161.5$ ml of 22 per cent Na_2SO_4 required.

Calculate the percentage concentration of 5 N H_2SO_4.

1. The equivalent weight of H_2SO_4 is $\dfrac{98}{2} = 49$; therefore, a 5 N solution contains $5 \times 49 = 245$ gm/l.

2. The $\dfrac{245 \text{ gm/l}}{10} = 24.5$ gm/100 ml = 24.5 per cent.

Another common conversion problem involves milliequivalents per liter. We have already seen that 1 ml of a 1 N solution contains 1 milliequivalent of a solute. Since electrolyte concentrations in body fuids are ordinarily expressed in milliequivalents per liter and many of the plasma constituents are represented in milligrams per 100 milliliters, conversion problems are involved. In general, to convert mg/100 ml to milliequivalents/l (mEq/l), we must first multiply by 10 to obtain mg/l and then divide by the equivalent weight to obtain mEq/l.

Examples

Convert 350 mg/100 ml of Na^+ to mEq/l.

1. The milligrams per liter = $350 \times 10 = 3500$ mg/l.

2. The mg/l divided by the equivalent weight of Na = $\dfrac{3500}{23} = 152.2$ mEq/l.

A solution of NaCl has a concentration of 145 mEq/l. Convert this to percentage concentration.

1. Since $\dfrac{mg/l}{equiv. \text{ wt.}} = $ mEq/l, then mg/l = mEq/l $\times$ equiv. wt. or mg/l = 145 $\times$ 58.5 = 8500 mg/l.

2. 8500 mg/l = 850 mg/100 ml = 0.85 gm/100 ml = 0.85 per cent NaCl

CONCENTRATION OF SOLUTIONS

EXERCISES

A. How many milliliters of 10 per cent HCl would be required to prepare 600 ml of a 0.5 M solution?

B. Calculate the normality of:

   a. Concentrated HCl (specific gravity 1.19; 37.5 per cent HCl by weight).

   b. Concentrated H_2SO_4 (specific gravity 1.84; 98 per cent H_2SO_4 by weight).

   c. 50 per cent NaOH solution.

ADDITIONAL EXERCISES

1. A liter of 5 per cent glucose solution contains what weight of glucose?

2. A 24 hour urine specimen (1250 ml) weighed 1275 gm. What is the specific gravity of the urine specimen?

3. Concentrated sulfuric acid has a specific gravity of 1.84 and is 98 per cent H_2SO_4 by weight. Calculate the weight of H_2SO_4 in 55.0 ml of concentrated sulfuric acid.

4. If you were offered 5 lbs of $CuSO_4 \cdot 5H_2O$ for $5.00 and 5 lbs. of anhydrous $CuSO_4$ for $7.00, which would be the best buy?

5. How many grams of H_2SO_4 would 1 liter of 1×10^{-3} M solution contain?

6. What weight of NaCl would be required to prepare 800 ml of 1 M solution?

7. A liter of 0.1 N H_2SO_4 was prepared from concentrated H_2SO_4 (specific gravity 1.84; 98 per cent by weight). How many milliliters of the concentrated acid was required?

8. Sixteen liters of approximately 0.1 N HCl were prepared from concentrated HCl. An accurate titration indicated a normality of 0.1037. How much water should be added to the 16 l to prepare 0.100 N HCl?

9. A stock solution of NaOH was exactly 12.0 N. How many milliliters of the stock solution would be used to prepare 16 l of 0.200 N NaOH?

10. Calculate the percentage concentration of 8 N HCl.

11. How many milliliters of 36.5 per cent HCl would be required to prepare 500 ml of a 5.0 N solution?

12. A patient's serum had a NaCl concentration of 609 mg/100 ml. What would the concentration of NaCl be in mEq/l?

CHAPTER 5

THE PREPARATION OF GRAPHS

In high school and college mathematics courses the subject of functional relationships and graphs of functions is almost always covered. You may recall the simplest type of functional relationship in which y is directly proportional to x. This may be expressed as a general equation:

$$y = ax$$

where a is a constant. The functional relationship that is relevant to the preparation of graphs in the laboratory is the linear function:

$$y = ax + b$$

where a and b are both constants. **You may recognize this as the equation of the graph for a straight line with a slope of a and an intercept of b.** In the preparation of graphs we very often choose an intercept of 0, or b = 0, and the resultant straight line represents a linear function of y with respect to ax, as shown in the first equation. Since methods in the laboratory are calibrated by the preparation of a standard curve, x may represent the concentration of the standard solutions and y the absorbance readings obtained when the standards are run through the method. To illustrate the plotting of a graph through a series of points that correspond to successive values of x and y we may assume that the following values of x represent blood glucose values in mg/100 ml and y the absorbance of the respective solutions.

X	0	50	100	150	200	250
Y	0	0.151	0.303	0.449	0.602	0.748

It is common practice in the laboratory to plot concentration or time along the x or horizontal axis (abscissa) and change in absorbance, O.D., %T or changes in concentration with time on the y or vertical axis (ordinate).

5.1 THE CHOICE OF UNITS FOR THE COORDINATES

For practical purposes such as reading blood level values from a calibration curve **it is essential to use most of the area on the graph paper** in plotting the graph. This requires inspection of the graph paper to determine the number of major squares along both the ordinate and abscissa. For example, the commonly used millimeter square graph paper (Dietzen No. 341-M) has 18 major squares along the ordinate and 25 major squares along the abscissa. Along the x axis we would then choose each 5 major squares to represent 50 mg/100 ml blood glucose. Each major square on the y axis could represent an absorbance of 0.050, covering 15 major squares compared to the 25 major squares utilized on the x axis. When the graph is constructed in this fashion and the points for each set of x, y values are plotted, a straight line covering most of the area of the graph paper results (Figure 5.1). The wrong choice of units on either the ordinate or abscissa, as illustrated in Figure 5.2, could result in a graph of the function that would not only be awkward to read but would also produce inaccurate values.

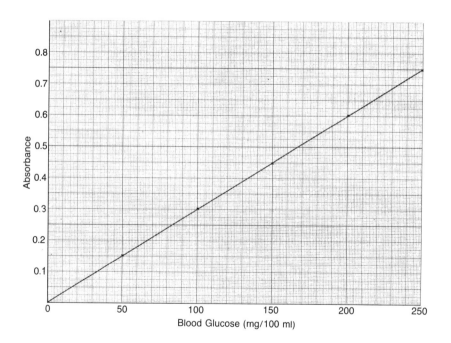

FIGURE 5.1

THE PREPARATION OF GRAPHS

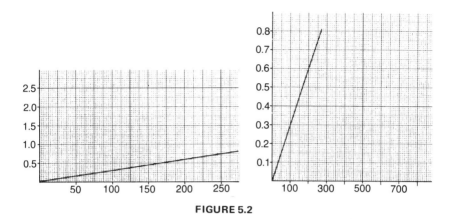

FIGURE 5.2

Another example may be the preparation of a calibration curve for serum inorganic phosphorus. Values in the following tabulation were obtained for absorbance versus calibration standards concentration in mg/100 ml.

Conc. mg/100 ml (X)	0	2.0	4.0	6.0	8.0
Absorbance (Y)	0	0.198	0.401	0.603	0.799

If millimeter square graph paper (No. 341-M) is used, each five major squares on the abscissa could represent 2.0 mg/100 ml P and each major square on the ordinate 0.050 absorbance. The straight line obtained by joining the plotted points would then cover 20 square on the abscissa and 16 squares on the ordinate, as shown in Figure 5.3.

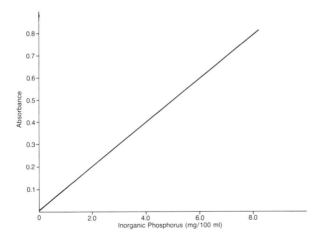

FIGURE 5.3

5.2 THE GLUCOSE TOLERANCE CURVE

Glucose tolerance tests are often run in a clinical laboratory. A blood specimen is drawn from a fasted patient and 75 to 100 grams of glucose is given by mouth, followed by the drawing of additional blood specimens at specific time intervals. A graph illustrating the glucose tolerance curve of the patient would be constructed by plotting changes in blood glucose with time. The results of such a tolerance test are shown in the following tabulation.

Time (Hours) (X)	0	0.5	1.0	1.5	2.0	2.5
Blood Glucose (mg/100 ml) (Y)	70	90	120	150	130	100

On millimeter square graph paper (No. 341-M) each five major squares on the abscissa could represent 0.5 hour and each five major squares on the ordinate 50 mg/100 ml blood glucose. The resultant curve would obviously not be a straight line and would not start at zero on the ordinate (Figure 5.4). It would follow the patient's blood sugar level starting at 70, rising to 150 at 1.5 hours, then dropping to 100 mg/100 ml at 2.5 hours after the ingestion of the glucose.

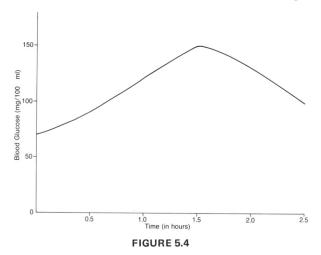

FIGURE 5.4

5.3 THE PREPARATION OF STANDARD CURVES

Although we have already discussed standard curves and used values from standard solutions of glucose and phosphorus to illustrate the graphs of linear functions, the subject is worth exploring in greater detail. When setting up a new or modified spectrophotometric method in the laboratory, the first step should involve **the construction of a calibration curve.** The curve should cover the entire

THE PREPARATION OF GRAPHS

range of concentrations met in practice, and it is essential to duplicate experimental conditions that will be employed in the routine analysis of specimens. Assuming that accurate standards are prepared and that the procedure is carried out with satisfactory technique, the spectrophotometric readings shown in Table 5.1 may result.

TABLE 5.1

Standard concentration (mg/100 ml)	% T	Absorbance	Klett readings
0	100.0	0.000	0.0
10	80.3	0.095	48.0
25	57.7	0.239	120.5
50	33.2	0.480	240.5
75	19.2	0.718	359.0
100	11.1	0.960	480.0

The three types of readings given in the above tabulation are obtained from instruments and scales commonly encountered in clinical laboratories. They are related to each other as follows:

$$\text{Absorbance} = 2 - \log \% \text{ T}$$

$$\text{Klett Readings} = \text{Absorbance} \times 500$$

At least four types of standard curves can be prepared from the above data plotting concentration on the abscissa versus spectrophotometric readings on the ordinate. Figure 5.5 illustrates a plot of % T (% transmission) versus concentration. This type of plot is not recommended for routine use for several reasons. The relationship is not linear, nor is it apparent from the curve whether the readings obtained from the method follows Beer's law (see Chapter 6, p. 65). Several concentration points must be obtained to construct a smooth curve, and errors of measurement are more difficult to assess. When % T values are plotted on one cycle semilog paper, a linear relationship between the readings and the concentration is evident, as seen in Figure 5.6. The direct and linear relationship between absorbance or Klett readings and the concentration is shown in Figures 5.7 and 5.8. **This type of plot is recommended for routine use in the laboratory.** It produces a straight line with a positive slope in which an increase in concentration results in an increase of absorbance or Klett readings. The relationship is linear, and a positive test of Beer's law is obtained from the readings. Once the standard curve is prepared for a method, it may be routinely checked by the use of a reagent blank plus two standard solutions, which represent the concentration of the highest standard and one-half the concentration of that standard in the accurate range of the method. In the standard curves shown in Figures 5.7 and 5.8, for example, the zero concentration or reagent blank, the 50 mg/100 ml and the 100 mg/100 ml standards would be used to check the calibration curve of the method.

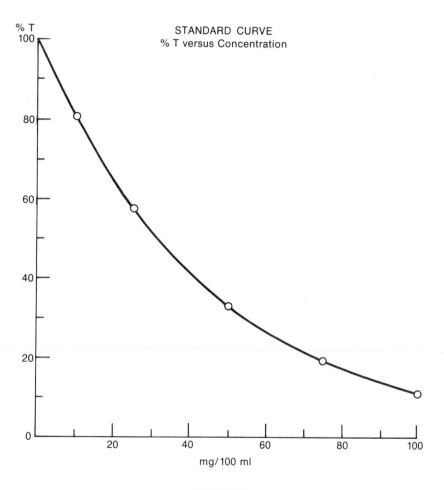

FIGURE 5.5

THE PREPARATION OF GRAPHS 61

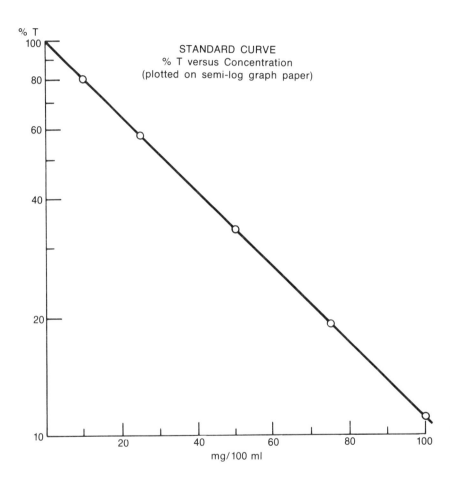

FIGURE 5.6

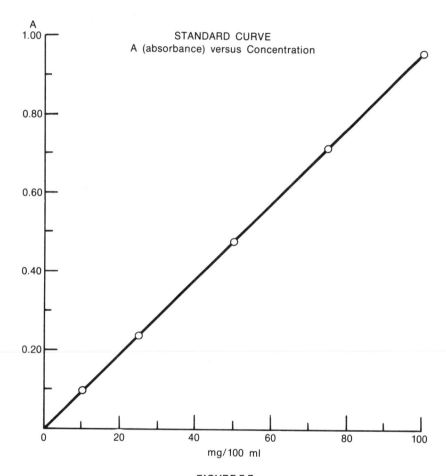

FIGURE 5.7

THE PREPARATION OF GRAPHS

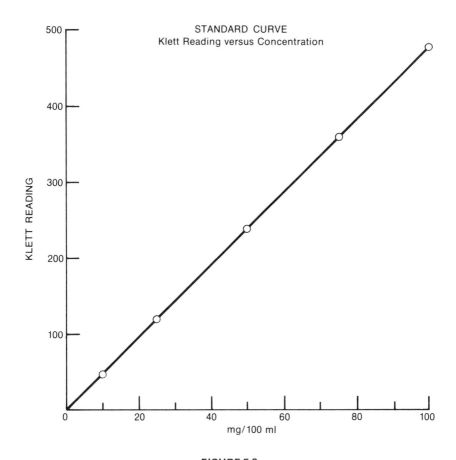

FIGURE 5.8

The current trend by laboratory instrument manufacturers is to produce **spectrophotometers with linear absorbance scales, digital displays or print-out systems.** These instruments will obviate the conversion of % T to absorbance values and simplify the preparation of linear calibration curves. For laboratories that use instruments containing only a % T scale the Table in the Appendix permits ready conversion of such readings to absorbance values.

EXERCISES

1. Explain the terms in the following equation: $y = ax + b$.

2. Which two terms in the above equation are most important in the clinical laboratory? Explain.

3. In the preparation of a calibration curve in the laboratory the following data was collected.

Conc. mg/100 ml	0	2.5	5.0	7.5	10.0
Absorbance	0	0.198	0.401	0.595	0.798

Plot the points and construct the standard curve from these values.

4. The only spectrophotometer in the laboratory had a scale graduated in % T. The readings obtained for a series of standard solutions were as follows:

Conc. mg/100 ml	0	5	10	25	50
% T		100.0	81.5	66.5	37.5	14.0

Plot the standard curve as absorbance versus concentration values.

5. A calibration curve was prepared in a laboratory employing a Klett photometer. The following data was obtained.

Conc. mg/100 ml	0	25	50	75	100
Klett readings	0.0	93.5	187.5	281.0	375.0

Plot the standard curve using both Klett readings and absorbance versus concentration values.

CHAPTER 6

SPECTROPHOTOMETRIC CALCULATIONS

The majority of the determinations carried out in the clinical chemistry laboratory depend on measurements made with a spectrophotometer. These instruments measure the amount of light absorbed by a particular substance in solution or a color produced by that substance in the visible range, the near ultraviolet and ultraviolet range and in the infrared range. **Calculations involving spectrophotometric data are based on Beer's law.** Beer's law states that the concentration of a substance is directly proportional to the amount of light absorbed or inversely proportional to the logarithm of the transmitted light. The mathematical equation representing Beer's law is as follows:

$$A = abc = \log \frac{100}{\% T} = 2 - \log \% T$$

where A = absorbance, a = absorptivity, b = light path of solution in centimeters, c = concentration of the substance in solution and $\% T$ = per cent transmittance.

Although some spectrophotometers have only a $\% T$ scale, the majority have both A and $\% T$ scales, and the modern trend is toward linear A scales. In the above equation when the concentration units are moles per liter, a is called the molar absorptivity or the **molar extinction coefficient** and is designated ϵ. In spectrophotometry ϵ is defined as the absorbance (optical density) of a one molar solution of 1 cm thickness. From the equation:

$$A = abc = \epsilon bc$$

when b = 1 cm and c = 1 molar, $A = \epsilon$

The measurement of the molar extinction coefficient is often used to determine the purity of a compound with a known molecular weight. An example is the determination of the purity of bilirubin in clinical laboratories. A joint commission of pathologists and clinical chemists recommended that bilirubin preparations having an ϵ between 59,100 and 62,300 at a wavelength of 453 mμ in chloroform at 25°C were suitable for standards.

Examples

A bilirubin solution in chloroform at 25°C measured at a wavelength of 453 mμ in a 1 cm light path cuvette gave an absorbance of 0.528. The concentration of the solution was 5 mg bilirubin per liter. Calculate the molar extinction coefficient.

From the equation $A = \epsilon bc$ or $\epsilon = \dfrac{A}{bc}$

$$c = 5 \text{ mg/l (mol. wt.} = 572)$$

c in moles/l = $\dfrac{0.005 \text{ gm/l}}{572 \text{ gm/l}}$ = 0.00000873 or 8.73×10^{-6} M

Then $\epsilon = \dfrac{A}{1 \times 8.73 \times 10^{-6} \text{ M}} = \dfrac{0.528}{8.73 \times 10^{-6} \text{ M}} = \dfrac{5.28 \times 10^{-1}}{8.73 \times 10^{-6}}$

$= 0.605 \times 10^{-1 \; - \; (-6)} = 0.605 \times 10^{5} \quad = 60,500$

A purified bilirubin preparation was found to have an extinction coefficient of 61,500 under the conditions described above. What concentration of this bilirubin expressed in mg/l would produce an absorbance of 0.850 at 453 mμ in a 1 cm light path cuvette at 25°C?

From the equation $A = \epsilon bc$, $c = \dfrac{A}{\epsilon b}$

$$c = \dfrac{0.850}{61,500 \times 1} = \dfrac{8.50 \times 10^{-1}}{6.15 \times 10^{4}} = 1.38 \times 10^{-1 \; - \; (4)}$$

$$= 1.38 \times 10^{-5}$$

Therefore $c = 1.38 \times 10^{-5}$ M $= \dfrac{X \text{ gm/l}}{572 \text{ gm/l}}$ or $X = 1.38 \times 10^{-5} \times 572$ or $X = 1.38 \times 10^{-5} \times 5.72 \times 10^{2} = 7.9 \times 10^{-3}$ gm/l = 7.9 mg/l.

SPECTROPHOTOMETRIC CALCULATIONS

EXERCISES

A. Calculate the molar extinction coefficient of a compound in a solution containing 0.1 mM per liter that produced an absorbance of 0.425 at 420 mμ in a 1 cm light path cuvette.

B. What absorbance value would result from reading a bilirubin solution, 1.0 mg%, ϵ = 62,000 in a 1 cm light path cuvette at 453 mμ at 25°C?

A more common type of calculation that is employed in spectrophotometric measurements involves the determination of the concentration of a known constituent in a biological fluid. From Beer's law the absorbance A of a solution of concentration C_1 is directly proportional to the absorbance A_2 of the same solution at a different concentration, C_2, or

$$\frac{A_1}{A_2} = \frac{C_1}{C_2}$$

$$\text{therefore } C_1 = \frac{A_1}{A_2} \times C_2$$

If solution 2 is a standard solution of known concentration and solution 1 a biological fluid of unknown concentration, then:

$$C_u = \frac{A_u}{A_s} \times C_s \text{ or in terms of \% T}$$

$$C_u = \frac{2 - \log \% T_u}{2 - \log \% T_s} \times C_s$$

Example

An example of a simple problem would be the calculation of the concentration of a constituent in urine when 1 ml of urine and 1 ml of a standard solution of the constituent containing 0.4 mg were treated by a reaction that produced an absorbance of 0.325 for the standard solution and 0.350 for the urine.

From the equation:

$$C_u = \frac{A_u}{A_s} \times C_s$$

$$C_u = \frac{0.350}{0.325} \times 0.4 \text{ mg} = 0.431 \text{ mg/ml of urine}$$

The same problem using % T values instead of absorbance would result in readings of 47.3% T for the standard solution and 44.6% T for the urine.

From the equation:

$$C_u = \frac{2 - \log \% T_u}{2 - \log \% T_s} \times C_s$$

$$2 - \log \% T_u = 2 - \log 44.6 = 2 - 1.65 = 0.35$$

$$2 - \log \% T_s = 2 - \log 47.3 = 2 - 1.675 = 0.325$$

$$C_u = \frac{0.350}{0.325} \times 0.4 \text{ mg} = 0.431 \text{ mg/ml urine}$$

When a dilution of serum, urine or a protein-free filtrate is used in a spectrophotometric determination or when the amount of a constituent in a definite volume of a biological fluid is desired, the calculation is more involved.

Examples

A patient excreted 1500 ml of urine in 24 hours. A 1 ml aliquot of the urine and 1 ml of a standard creatinine solution containing 0.8 mg were treated with an alkaline picrate solution with a resulting absorbance of 0.285 for the standard and 0.275 for the urine when read at 530 mμ. Calculate the 24 hr excretion of creatinine.

From the equation:

$$C_u = \frac{A_u}{A_s} \times C_s \; ; \; C_u = \frac{0.275}{0.285} \times 0.8 \text{ mg} = 0.772 \text{ mg/ml}$$

Total volume: 1500 ml $\times$ 0.772 mg/ml = 1158 mg/24 hr

SPECTROPHOTOMETRIC CALCULATIONS

A 1 ml aliquot of a 1:10 protein-free filtrate from serum and 0.2 ml of a standard phosphate solution containing 0.003 mg P diluted to a volume of 1 ml with water were treated with an aminonaphtholsulfonic acid solution to produce a blue color that was diluted to 25 ml in each determination; the absorbancies were read at 660 mμ. The reading of the standard solution was 0.410 and of the protein-free filtrate 0.450. Calculate the phosphate concentration in mg% in the serum.

From the equation:

$$C_u = \frac{A_u}{A_s} \times C_s \; ; \; C_u = \frac{0.450}{0.410} \times 0.003 \text{ mg} = 0.00329 \text{ mg}$$

$$C_u = 0.00329 \text{ mg in } \frac{1}{10} \text{ ml serum} = 10 \times 0.00329 = 0.0329 \text{ mg/ml}$$

or 100×0.0329 mg = 3.29 mg% P in serum

ADDITIONAL EXERCISES

1. A vitamin D solution in alcohol at 25°C measured at a wavelength of 264 mμ in a 1 cm light path cuvette gave an absorbance of 0.218. The concentration of the solution was 5 mg/l and the molecular weight is 397. Calculate the molar extinction coefficient.

2. Using the molar extinction coefficient found in problem 1 for vitamin D, calculate the absorbance at 264 mμ in a 1 cm light path cuvette at 25°C produced by a solution whose concentration was 12.5 mg/l.

3. A 1 ml aliquot of urine and 1 ml of a standard creatinine solution containing 0.4 mg were treated with an alkaline picrate solution with a resulting absorbance of 0.582 for the standard and 0.560 for the urine when read at 530 mμ. Calculate the creatinine excretion in mg/100 ml.

4. To a 4 ml aliquot of a 1:10 protein free filtrate from serum and 4 ml of a standard creatinine solution containing 0.00115 mg/ml were added 2 ml of an alkaline picrate solution. After 15 minutes the absorbance readings at 530 mμ were as follows: protein free filtrate aliquot 0.365, standard 0.350. What was the creatinine concentration of the serum in mg/100 ml?

5. A series of standard phosphate solutions was treated with an amino-naphtholsulfonic acid solution to produce a blue color that was diluted to 25 ml in each determination and the absorbancies were read at 660 mμ.

Phosphate equiv. to mg%	Absorbance
0	0.003
1.0	0.150
2.0	0.295
3.0	0.459
4.0	0.595
5.0	0.725

Construct the standard curve plotting the concentration (abscissa) against the absorbance (ordinate). Does the curve follow Beer's law throughout the entire range of concentrations? Explain.

6. From the standard curve prepared in Problem 5, determine the serum phosphate concentrations of the following specimens that were treated the same as the standards.

Absorbance readings
a. 0.625
b. 0.505
c. 0.495
d. 0.640

7. A flame emission spectrophotometer set up for the determination of serum sodium gave readings of 19, 40, 59, 81 and 99 on a linear scale when standards containing 110, 120, 130, 140 and 150 mEq/l of sodium respectively were aspirated in the flame. What would be the sodium concentration of the following serum samples that were treated in the same fashion as the standards?

Meter readings
a. 51
b. 83
c. 69
d. 72

CHAPTER 7

QUALITY CONTROL STATISTICS

Quality control methods are commonly employed in the modern clinical laboratory to ensure the constant production of reliable results. There are many expressions, borrowed from statistics, that are used to maintain confidence in analytic values. A comparison, for example, of the meaning of the terms accuracy, precision and reliability may help set the stage for a discussion of the application of statistics to laboratory mathematics.

The **accuracy** of a method for the determination of a substance refers to the agreement of the results with the true value for that substance in the specimen. In the clinical laboratory it is not always possible to obtain primary standards that represent the true value of a constituent, and reference standards are often used. The **precision** of a method refers to its ability to produce a series of results that agree very closely with each other. If ten determinations of a single constituent result in values that are in close agreement, the method has a high degree of precision. This obviously does not mean that the method is necessarily accurate since there are many examples of methods for glucose, urea and calcium that exhibit precision and yet do not measure the true value for the particular constituent. The **reliability** of a method is a measure of its ability to achieve both accuracy and precision. A reliable method would therefore be one that consistently produced accurate and precise results in the determination of a laboratory constituent.

Blood glucose values obtained by the use of tungstic acid protein-free filtrates (Folin-Wu) and by a glucose oxidase "true glucose" method may be used to illustrate statistical calculations. Ten determinations were run by each method on the same specimen and the values are shown in Table 7.1. The **mean value** for each method, $\bar{X}$, equals the sum of the ten values divided by N, the number of individual results or:

$$\text{Mean} = \bar{X} = \frac{\Sigma x}{N}$$

TABLE 7.1

Folin-Wu glucose mg/100 ml	Deviation from mean	Deviation squared	True glucose mg/100 ml	Deviation from mean	Deviation squared
98	−1	1	87	+1	1
101	+2	4	85	−1	1
99	0	0	86	0	0
97	−2	4	83	−3	9
98	−1	1	89	+3	9
102	+3	9	84	−2	4
99	0	0	86	0	0
101	+2	4	87	+1	1
98	−1	1	85	−1	1
97	−2	4	88	+2	4
990		28	860		30

Mean = $\bar{X} = \dfrac{\Sigma X}{N} = \dfrac{990}{10} = 99$ mg/100 ml   Mean = $\bar{X} = \dfrac{\Sigma X}{N} = \dfrac{860}{10}$ 86 mg/100 ml

Range 97 – 102 mg/100 ml   Range 83 – 89 mg/100 ml

$$\text{Standard deviation } s = \sqrt{\dfrac{\Sigma (X - \bar{X})^2}{N - 1}}$$

$s = \sqrt{\dfrac{28}{9}} = \sqrt{3.11} = 1.76$ mg/100 ml $s = \sqrt{\dfrac{30}{9}} = \sqrt{3.33} = 1.83$ mg/100 ml

$$\text{C. V.} = \dfrac{s}{\bar{X}} \times 100$$

C. V. = $\dfrac{1.76}{99} \times 100 = 1.78$ per cent   C. V. = $\dfrac{1.83}{86} \times 100 = 2.13$ per cent

QUALITY CONTROL STATISTICS

The **range of values** is an expression of the precision or repeatability of the method. A very commonly used measure of precision is called the standard deviation, s, which represents the scattering of values around the mean and is related to distribution curves and quality control charts. The **standard deviation** equals the square root of the sum of the squared differences from the mean divided by one less than the number of determinations.

$$s = \sqrt{\frac{\Sigma (X - \bar{X})^2}{N - 1}}$$

Another method that is used to compare the results of two sets of data is the calculation of the **coefficient of variation**. This term expresses the scattering of results as percentages of the mean and may be used to compare values of different magnitudes and different units. The coefficient of variation, C. V., may be expressed as follows:

$$C.V. = \frac{S}{\bar{X}} \times 100$$

Calculations for these expressions are illustrated for the two sets of blood glucose values in Table 7.1.

From the calculated results for the two methods we would conclude that the Folin-Wu type method demonstrated the most precision, i.e., lower standard deviation and coefficient of variation, whereas the true glucose method was more accurate since the values grouped around the true glucose value.

EXERCISES

A. A serum specimen was analyzed ten times for its chloride content. The following values, expressed in milliequivalents/liter, were obtained: 106, 103, 105, 102, 105, 104, 106, 103, 102, 104. Calculate the standard deviation.

B. A serum specimen was obtained from a male laboratory technician and a female laboratory technician. Each specimen was analyzed ten times for uric acid content with the following results: Male 6.2, 6.0, 6.0, 6.4, 5.9, 6.1, 6.4, 5.8, 6.3, 5.9. Female 5.5, 5.7, 5.9, 5.7, 5.6, 5.8, 5.6, 5.6, 5.9, 5.7. Calculate the standard deviation and coefficient of variation for each set of values.

In quality control systems the deviation of a value from the true value and the distribution of the values around the mean are important considerations. If a single constituent is determined in the serum of a large group of individuals and the **frequency of occurrence** of each value is plotted, the result is a normal or gaussian distribution curve (Figure 7.1).

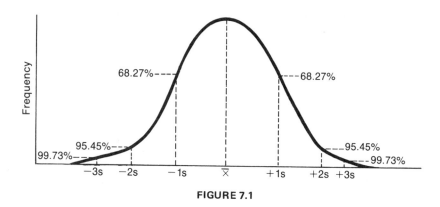

FIGURE 7.1

The mean value corresponds to the peak in the distribution of results in a **normal distribution curve.** If the standard deviation is calculated for a large series of results the percentage distribution percentages would be as follows:

± 1.0 standard deviation from the mean 68.27%
± 2.0 standard deviation from the mean 95.45%
± 3.0 standard deviation from the mean 99.73%

A **quality control chart** usually represents the distribution of values of a reference standard as follows:

```
+3s  ----------------------------------------
+2s  ----------------------------------------
+1s  ----------------------------------------
X̄    _____
-1s  ----------------------------------------
-2s  ----------------------------------------
-3s  ----------------------------------------
```

QUALITY CONTROL STATISTICS

As an example, we may represent the values that were calculated for the true glucose method on such a chart. The mean value was 86 mg/100 ml, and s was 1.8 mg/100 ml.

```
91.4  _ _ _ _ _ _ _ _ _ _ _ _ _ _ _ _ _ _ _ _ _ _ _ _ _ _ _ _   +3s
89.6  _ _ _ _ _ _ _ _ _ _ _ _ _ _ _ _ _ _ _ _ _ _ _ _ _ _ _ _   +2s
                              •
87.8  _ _ _ _ _ _ _ _ _ _ _ _ _ _ _ _ _ _ _ •_ _ _ _ •_ _ _ _   +1s
           •
86.0  ─────────────────•─────────────────────•────────────────  Mean
                                                       •
84.2  _ _ _ _ _ •_ _ _ _ _ _ _ _ _ _ _ _ _ _ _ _ _ _ _ _ _ _    −1s
                                •
82.4  _ _ _ _ _ _ _ _ _ _ _•_ _ _ _ _ _ _ _ _ _ _ _ _ _ _ _ _   −2s
80.6  _ _ _ _ _ _ _ _ _ _ _ _ _ _ _ _ _ _ _ _ _ _ _ _ _ _ _ _   −3s
```

Glucose mg/100 ml

In most laboratories the acceptable limits of precision for a method used in quality control is ± 2.0s, which includes 95.5 per cent of all the values obtained for the reference standard.

ADDITIONAL EXERCISES

1. Construct a quality control chart for the uric acid values for the male laboratory technician (Exercise B, p. 73) including ± 2.0 standard deviations as the limit of control.

2. A pooled serum sample was analyzed 20 times for its urea nitrogen content. Values in mg/100 ml were obtained as follows: 12.2, 12.4, 12.0, 12.5, 12.3, 12.3, 12.6, 12.0, 12.5, 12.2, 12.4, 12.1, 12.6, 12.2, 12.3, 12.2, 12.5, 12.4, 12.4, 12.6. Construct a quality control chart from these values including ± 2.0 standard deviations from the mean.

3. Construct a quality control chart for the Folin-Wu glucose values in Table 7.1. Plot the ten values and indicate those which fall outside the ±2.0 s limits.

CHAPTER 8

HYDROGEN ION CONCENTRATION AND pH

In biological fluids, which are essentially water solutions, an acid may be defined as a substance that yields protons and a base as a substance that accepts protons. From a more practical standpoint, in an aqueous solution an **acid** may be regarded as a substance that **yields hydrogen ions** ($\overset{+}{H}$) and a **base** as a substance that **yields hydroxide ions** ($\overline{OH}$). When an acid or base is added to water, the degree of dissociation into ions will determine the relative strength of the acid or basic solution. In a dilute solution a **strong acid** such as HCl will dissociate almost completely into ($\overset{+}{H}$) and ($\overline{Cl}$) ions. A **weak acid,** such as acetic acid, under the same conditions will dissociate only slightly into ($\overset{+}{H}$) and ($CH_3 \overline{COO}$) ions. Assuming 100 per cent ionization of HCl and 1 per cent ionization of acetic acid in 0.1 N solution, the [$\overset{+}{H}$] concentration could be calculated as follows:

$$N \times \% \text{ ionization} = [\overset{+}{H}] \text{ concentration}$$

For HCl $0.1 \text{ N} \times 1.00 \ (100\%) = 0.100 = [\overset{+}{H}]$
For acetic acid $0.1 \text{ N} \times 0.01 \ \ (1\%) = 0.001 = [\overset{+}{H}]$

Water dissociates very slightly into ($\overset{+}{H}$) and ($\overline{OH}$) ions as follows:

$$H_2O \rightleftarrows \overset{+}{H} + \overline{OH}$$

It can be shown that 10,000,000 liters of water will yield 1 g of ($\overset{+}{H}$), or that one liter of water contains 1/10,000,000 of a gram of hydrogen ions. The **hydrogen ion concentration** of water therefore equals $\frac{1}{10^7}$ or 10^{-7} g/l. The hydrogen ion and hydroxide ion concentrations of water must be equal since water is a neutral solution. The [$\overset{+}{H}$] concentration = 10^{-7} and the [$\overline{OH}$] concentration = 10^{-7} and the dissociation constant, K_W, for water equals

$$K_W = 10^{-7} \times 10^{-7} = 10^{-14}$$

HYDROGEN ION CONCENTRATION AND pH

Since a neutral solution such as water has a hydrogen ion concentration of 10^{-7}, it follows that an acid solution must have a $[\overset{+}{H}]$ concentration greater than 10^{-7} and that a basic solution must have a $[\overset{+}{H}]$ concentration less than 10^{-7}.

In 1909, Sorensen decided that laboratory workers would experience difficulty in expressing $[\overset{+}{H}]$ concentrations of acid and basic solutions in terms of exponential numbers. He devised the term **pH**, which he defined as **the negative logarithm of the $[\overset{+}{H}]$ concentration.** The relation is exceedingly simple if the $[\overset{+}{H}]$ concentration is exactly 1×10^{-power} because the pH then equals the value of the minus power of ten. For example, water has a hydrogen ion concentration of 10^{-7}, or a pH of 7; the 0.1 N HCl from the example above has a $[\overset{+}{H}]$ concentration of 0.1, or 10^{-1}, or a pH of 1; and the 0.1 N acetic acid with a $[\overset{+}{H}]$ concentration of 0.001, or 10^{-3}, has a pH of 3. To calculate the pH of a solution with a $[\overset{+}{H}]$ concentration more complex than 1×10^{-power} we must resort to the use of logarithms as follows:

$$pH = \log \frac{1}{[\overset{+}{H}]} = -\log [\overset{+}{H}]$$

Since the $[\overset{+}{H}]$ is expressed as a $\times 10^{-b}$ the pH = $-\log a \times 10^{-b}$ or pH = b $-$ log a. If the $[\overset{+}{H}]$ concentration of a solution is 0.004, calculate the pH.

Example

$$\text{The } [\overset{+}{H}] = a \times 10^{-b} = 4 \times 10^{-3}$$

$$\text{The pH} = -\log 4 \times 10^{-3} \text{ or } 3 - \log 4 \text{ as } (b - \log a)$$

$$\text{Calculating, } 3 - \log 4 = 3 - 0.602 = 2.398 = pH$$

The pH value for any $[\overset{+}{H}]$ concentration can be approximated mentally by expressing the $[\overset{+}{H}]$ concentration between 1 and 10 $\times$ 10 to a minus power. For example, 0.005 can be expressed as 5×10^{-3} and falls between 1×10^{-3} and 10×10^{-3}. The pH corresponding to $1 \times 10^{-3} = 3$ and to $10 \times 10^{-3} = 1 \times 10^{-2} = 2$. If the logarithm scale were linear, we would estimate 5×10^{-3} as a pH of 2.5, however a quick study of the approximate logarithms of values from 1 to 9 will yield a more accurate estimate. This method of approximation is illustrated in the following tabulation:

TABLE 8.1

Number × 10^{-power} (a × 10^{-b})	Approximate Logarithm	$(\overset{+}{H})$ Concentration	Approximate pH (b − log a)
1	0	1×10^{-3}	3.0
2	0.3	2×10^{-3}	2.7
3	0.5	3×10^{-3}	2.5
4	0.6	4×10^{-3}	2.4
5	0.7	5×10^{-3}	2.3
6	0.8	6×10^{-3}	2.2
7	0.85	7×10^{-3}	2.2
8	0.9	8×10^{-3}	2.1
9	0.95	9×10^{-3}	2.0

The use of this approximation would result in a more accurate value for the pH of a solution with a $[\overset{+}{H}]$ concentration of 5×10^{-3}. The value of 2.3 compares favorably with the value obtained by use of a logarithm table.

If the pH of a solution is known and the $[\overset{+}{H}]$ is desired, the calculation may be carried out as follows:

If pH = 4.5 and pH = − log $[\overset{+}{H}]$ = 4.5,

then   log $[\overset{+}{H}]$ = − 4.5 = (−5 + 0.5) = 0.5 − 5

Therefore: $[\overset{+}{H}]$ = antilog (0.5 − 5) = (antilog 0.5) × (antilog − 5)
$[\overset{+}{H}]$ = 3.16 × 10⁻⁵

EXERCISES

1. A 0.01 N acetic acid solution is 3 per cent ionized. Calculate the $[\overset{+}{H}]$ concentration.

2. A 0.1 N HCl solution is 95 per cent ionized. Calculate the $[\overset{+}{H}]$ concentration.

3. An acid solution has 1/10,000 gm of $(\overset{+}{H})$ per liter. Express the $[\overset{+}{H}]$ concentration in terms that would permit you to mentally calculate the pH. What is the pH?

HYDROGEN ION CONCENTRATION AND pH

4. Mentally estimate the pH of a solution with a $[\overset{+}{H}]$ concentration of 0.0004 (4×10^{-4}).

5. A 0.02 N H_3PO_4 solution is 70 per cent ionized. Calculate the $[\overset{+}{H}]$ concentration. Mentally estimate the pH and then calculate the actual pH.

6. A biological fluid had a $[\overset{+}{H}]$ concentration of 0.0000000312. What is the pH of the fluid?

7. An acid solution had a pH of 3.5. Calculate its $[\overset{+}{H}]$ concentration.

CHAPTER 9

BUFFERS

In the last chapter we discussed the dissociation of water and acids and the calculation of the hydrogen ion concentration and pH. As you doubtless know, **buffer systems consist of a mixture of a weak acid and its salt** and are capable of resisting, or buffering, additions of acids or bases. The ionization constant for a weak acid, such as acetic acid, is equal to the product of the ionic concentrations divided by the concentration of the undissociated acid. All concentrations are usually expressed in moles per liter but may also be expressed in millimoles or milliequivalents per liter. The ionization constant of a weak acid is usually designated as K_a, and we may write the equilibrium equation for acetic acid as follows:

$$HC_2H_3O_2 \rightleftarrows \overset{+}{H} + C_2H_3O_2^-$$

$$K_a = \frac{[\overset{+}{H}] \times [C_2H_3O_2^-]}{[HC_2H_3O_2]}$$

Even though any weak acid and weak base can be used to prepare a buffer solution, most buffers are composed of a weak acid and weak base that are conjugates of each other. The acetate buffer, for example, employs acetic acid as the weak acid and acetate ion as the weak base. The preparation of such a buffer may involve the direct addition of acetic acid and an acetate salt to water, or the partial neutralization of an acetic acid solution by a strong base, such as sodium hydroxide, or the partial neutralization of an acetate salt solution by a strong acid, such as hydrochloric acid. The ratio of acetic acid to acetate ion in solution may be chosen to give the desired $[\overset{+}{H}]$ or pH for the buffer solution. If we rearrange the equilibrium equation of acetic acid:

$$K_a = \frac{[\overset{+}{H}] \times [C_2H_3O_2^-]}{[HC_2H_3O_2]} \text{ to } [\overset{+}{H}] = K_a \times \frac{[HC_2H_3O_2]}{[C_2H_3O_2^-]}$$

BUFFERS

The $[\overset{+}{H}]$ of the buffer could be calculated from the values for K_a and the concentrations of acid and acetate ion. To convert to the pH and pK notation we could first express the equation in logarithmic form as follows:

$$\log [\overset{+}{H}] = \log K_a + \log \frac{[HC_2H_3O_2]}{[C_2H_3O_2^-]}$$

Since pH and $pK_a = -\log [\overset{+}{H}]$ and $-\log K_a$, the equation could be written

$$-\log [\overset{+}{H}] = -\log K_a - \log \frac{[HC_2H_3O_2]}{[C_2H_3O_2^-]}$$

or as $$pH = pK_a - \log \frac{[HC_2H_3O_2]}{[C_2H_3O_2^-]}$$

Finally to retain a positive log function:

$$pH = pK_a + \log \frac{[C_2H_3O_2^-]}{[HC_2H_3O_2]}$$

If we substitute [salt] for $[C_2H_3O_2^-]$ and [acid] for $[HC_2H_3O_2]$, we obtain the **Henderson-Hasselbalch equation**, which is used for the calculation of buffers in general and of acid-base problems involving body fluids in particular.

$$pH = pK + \log \frac{[salt]}{[acid]}$$

The **maximum buffer capacity** of a buffer *in vitro* is obtained when equal concentrations of salt and acid are present. Under these conditions, as, for example, when the acid is half neutralized in a titration, the equation may be expressed as:

$$pH = pK + \log \frac{1}{1} = pK + 0 = pK$$

This represents a practical method of determining **the pK of a buffer**. Values for the pK of acids and bases may also be found in tables in appropriate handbooks.

Examples

The pK_a values for acetic acid and carbonic acid are 4.73 and 6.10, respectively. Calculate the dissociation constant for each acid. Which is the stronger acid?

Since $pK_a = -\log K_a$, $\log K_a = -pK_a$

For acetic acid

$$\log K_a = -4.73 = (-5 + 0.26) = 0.26 - 5$$
$$K_a = \text{antilog } 0.26 \times \text{antilog} - 5$$
$$= 1.8 \times 10^{-5}$$

For carbonic acid

$$pK_a = -\log 6.10$$
$$\log K_a = -6.10 = (-7 + 0.9) = 0.9 - 7$$
$$K_a = \text{antilog } 0.9 \times \text{antilog} - 7$$
$$= 7.95 \times 10^{-7}$$

From the K_a values, acetic acid $K_a = 1.8 \times 10^{-5}$ is more completely dissociated and is a stronger acid than carbonic acid, $K_a = 7.9 \times 10^{-7}$.

If 10 ml of 0.05 M acid, when mixed with 10 ml of 0.1 M solution of its sodium salt, gave a solution with a pH of 5.03, what is the pK_a of the acid? The total volume of the final solution is 20 ml; therefore, the final concentration of acid is 10/20 × 0.05 M = 0.025 M and the final concentration of the salt is 10/20 × 0.1 M = 0.05 M. The fraction:

$$\frac{[\text{salt}]}{[\text{acid}]} = \frac{0.05}{0.025} = 2.0$$

and from the Henderson-Hasselbalch equation:

$$pH = pK_a + \log \frac{[\text{salt}]}{[\text{acid}]} \quad ; \quad 5.03 = pK_a + \log 2$$

or $5.03 = pK_a + 0.3$ and $pK_a = 5.03 - 0.3 = 4.73$

What amounts (in grams) of sodium acetate and acetic acid are required to prepare 1 liter of 0.1 M buffer with a pH of 5.0? The pK is 4.73. From the Henderson-Hasselbalch equation:

$$pH = pK + \log \frac{[\text{salt}]}{[\text{acid}]} \quad ; 5.0 = 4.73 + \log \frac{[\text{salt}]}{[\text{acid}]}$$

Since the total acetate concentration is to be 0.1 M:

$$\text{moles salt/l} + \text{moles acid/l} = 0.1 \text{ moles/l}.$$

BUFFERS

From the equation $\log \frac{[salt]}{[acid]} = 5.0 - 4.73 = 0.27$

$$\frac{[salt]}{[acid]} = \text{antilog of } 0.27 = 1.862$$

Since salt + acid = 0.1 moles/l

$$\text{moles salt/l} = 0.1 - \text{moles acid/l}$$

and substituting 1.862 moles acid/l for moles salt/l

$$1.862 \text{ moles acid/l} = 0.1 - \text{moles acid/l}$$

$$\text{or } 2.862 \text{ moles acid/l} = 0.1$$

and moles acid/l = $\frac{0.1}{2.862}$ = 0.0349 moles acetic acid/l

then moles salt/l = 0.1 − 0.0349 = 0.0651 moles sodium acetate/l.

Weight of acetic acid per liter = 0.0349 × 60 (mol. wt.) = 2.094 gm.
Weight of sodium acetate per liter = 0.0651 × 82 (mol. wt.) = 5.345 gm.

Therefore, 5.345 gm of sodium acetate and 2.094 gm of acetic acid when dissolved and diluted to 1 liter will produce a 0.1 M acetate buffer with a pH of 5.0.

Given a buffer solution composed of 10 ml of 0.2 M sodium acetate and 10 ml of 0.2 M acetic acid; what pH change in the solution would result from the addition of 1 ml of 0.2 M HCl? The pK is 4.73. From the Henderson-Hasselbalch equation:

$$pH = pK + \log \frac{[salt]}{[acid]}$$

The original pH would be: $pH = pK + \log \frac{0.1 \text{ M}}{0.1 \text{ M}}$

$$pH = 4.73 + \log \frac{0.1}{0.1} = 4.73$$

Since the concentrations of the three solutions are equimolar, we could substitute volumes for concentrations as:

$$pH = 4.73 + \log \frac{10}{10} = 4.73$$

then after the addition of the HCl the equation would be:

$$pH = 4.73 + \log \frac{9}{11} = 4.73 + \log 9 - \log 11$$

$$= 4.73 + 0.954 - 1.041 = 4.64$$

The ratio changes from 10/10 to 9/11 since the HCl would convert 1 ml of the salt to 1 ml of the acid.

EXERCISES

A. Calculate the pH of a bicarbonate buffer that is composed of a solution with a final concentration of 0.50 M sodium bicarbonate and 0.05 M carbonic acid. The pK is 6.10.

B. If 16.4 gm of sodium acetate and 1.2 gm of acetic acid are dissolved and made to volume in a 1 liter volumetric flask, what is the resultant pH of the buffer? The pK is 4.73, the molecular weight of sodium acetate is 82 and the molecular weight of acetic acid is 60.

C. If 20 ml of 0.1 M acid, when mixed with 20 ml of 0.3 M solution of its sodium salt, gave a solution with a pH of 6.58, what is the pK_a of the acid?

D. Given a buffer solution composed of 100 ml of 0.1 M sodium acetate and 100 ml of 0.1 M acetic acid, what pH change would result from the addition of 5.0 ml of 0.1 M HCl. The pK_a is 4.73. (Hint: the concentrations are all 0.1 M.)

9.1 ACID-BASE BUFFER CALCULATIONS

In addition to hemoglobin, the $\frac{HCO_3^-}{H_2CO_3}$ buffer system is primarily responsible for the **transport of CO_2 and the control of blood pH**. The Henderson-

Hasselbalch equation representing this important blood buffer was originally written as:

$$pH = pK_a + \log \frac{[HCO_3^-]}{[H_2CO_3]}$$

The pK_a for the buffer is 6.1 and normal values for $[HCO_3^-]$ and $[H_2CO_3]$ average 27 and 1.35 mM/l, respectively. Under normal conditions the pH of the blood averages 7.4; substituting these values in the equation:

$$7.4 = 6.1 + \log \frac{27 \text{ mM/l}}{1.35 \text{ mM/l}}$$

or

$$7.4 = 6.1 + \log 20 = 6.1 + 1.3 = 7.4$$

The normal ratio of bicarbonate to carbonic acid of 20:1 is maintained by the buffer with the assistance of the lungs and the kidneys. Changes in this ratio are seen in respiratory and metabolic acid-base imbalances. For example, an initial change in bicarbonate concentration to one half the normal value without any change in carbonic acid concentration occurred in a rapidly developing metabolic acidosis. What change would occur in the blood pH?

From the Henderson-Hasselbalch equation:

$$pH = pK + \log \frac{[HCO_3^-]}{[H_2CO_3]}$$

$$= 6.1 + \log \frac{13.5 \text{ mM/l}}{1.35 \text{ mM/l}}$$

$$= 6.1 + \log 10 = 6.1 + 1.0 = 7.1$$

A major problem that concerns the use of the Henderson-Hasselbalch equation in the above form is the fact that free carbonic acid is not readily measured. It is, however, in equilibrium with and proportional to the dissolved CO_2, which is also directly proportional to **the partial pressure of CO_2 (pCO_2)**. These statements may be illustrated as follows:

$$H_2CO_3 \rightleftarrows CO_2 + H_2O \rightleftarrows \alpha p\, CO_2 \text{ (dissolved } CO_2\text{)} \rightleftarrows p\, CO_2$$

If we substitute $\alpha p\, CO_2$ for $[H_2CO_3]$

$$pH = pK + \log \frac{[HCO_3^-]}{\alpha p\, CO_2}$$

Since the total CO_2 of the blood = $[HCO_3^-] + [H_2CO_3]$, the $[HCO_3^-]$ = total $CO_2 - [H_2CO_3]$ or total $CO_2 - \alpha p\,CO_2$. Substituting again:

$$pH = pK + \log\left[\frac{\text{Total } CO_2 - \alpha p\,CO_2}{\alpha p\,CO_2}\right]$$

The solubility coefficient, α of CO_2 = 0.03, so using known values

$$pH = 6.1 + \log\left[\frac{\text{Total } CO_2 - 0.03\,p\,CO_2}{0.03\,p\,CO_2}\right]$$

When total CO_2 is expressed in mM/l and pCO_2 as mm Hg, the expression for pCO_2 becomes:

$$pCO_2 \text{ (mm Hg)} = \frac{\text{Total } CO_2 \text{ mM/l}}{0.03\,[\text{antilog (pH} - 6.1) + 1]}$$

Example

What is the pCO_2 of plasma with total CO_2 of 28 mM/l and a pH of 7.50?

From the equation:

$$pCO_2 = \frac{\text{Total } CO_2}{0.03\,[\text{antilog (pH} - 6.1) + 1]}$$

$$= \frac{28}{0.03\,[\text{antilog }(7.5 - 6.1) + 1]}$$

$$= \frac{28}{0.03\,[\text{antilog }(1.4) + 1]} = \frac{28}{0.03\,[25.1 + 1]}$$

$$= \frac{28}{0.03\,[26.1]} = \frac{28}{0.783} = 35.8$$

Therefore the pCO_2 = 35.8 mm Hg

Since the equation just employed is rather complex for routine calculations in the laboratory, several schemes have been devised to simplify the problem. These include **nomograms**, which illustrate the relationships existing in the Henderson-Hasselbalch equation which permit the determination of pCO_2 graphically if the total CO_2 and the pH are known. In addition **special slide rules**

BUFFERS

have been devised to facilitate calculations involving the equation. Another method requires a total CO_2 content determination of anaerobically separated plasma plus a determination of whole blood or "true" plasma pH at 38°C. An equation is then set up as follows:

$$pCO_2 \text{ (mm Hg)} = \frac{\text{Total } CO_2 \text{ mM/l}}{F}$$

where F is a factor that corresponds to the denominator of the earlier equation 0.03 [antilog (pH − 6.1) + 1] and is calculated for each pH value at a temperature of 38°C. A portion of an available table of factor F versus pH at 38°C is reproduced to illustrate the method.

TABLE 9.1

pH	Factor F	pH	Factor F	pH	Factor F
7.10	0.331	7.26	0.465	7.42	0.659
7.12	0.345	7.28	0.486	7.44	0.689
7.14	0.360	7.30	0.507	7.46	0.720
7.16	0.376	7.32	0.530	7.48	0.752
7.18	0.392	7.34	0.553	7.50	0.786
7.20	0.409	7.36	0.578	7.52	0.822
7.22	0.427	7.38	0.604	7.54	0.859
7.24	0.445	7.40	0.631	7.56	0.898

As a sample calculation we may determine the pCO_2 of plasma with a total CO_2 of 28 mM/l and a pH of 7.50, as used in the last example.

From the equation:

$$pCO_2 = \frac{\text{Total } CO_2}{F}$$

F from the table = 0.786; therefore,

$$pCO_2 = \frac{28}{0.786} = 35.6 \text{ mm Hg}$$

This value for pCO_2 compares favorably with the 35.8 mm Hg obtained with the full equation.

ADDITIONAL EXERCISES

1. A patient with the symptoms of metabolic acidosis had plasma values of 20 mM/l and 1.5 mM/l for bicarbonate and carbonic acid concentrations, respectively. What was the pH of the plasma?

2. Calculate the pCO_2 of a patient's plasma if the pH is 7.33 and the total CO_2 is 24 mM/l.

3. Using the table of factors, calculate the pCO_2 if the total CO_2 of the plasma is 37 mM/l and the pH is 7.56.

4. If a patient's plasma has a pH of 7.10 and a pCO_2 of 78 mm Hg, calculate the total CO_2 content in mM/l, using the table of factors.

5. What is the pH of the plasma of a patient whose bicarbonate and carbonic acid concentrations are 35 mM/l and 1.25 mM/l, respectively?

6. What value for plasma bicarbonate would be obtained on analysis of a serum specimen with the following characteristics: pH 7.20 and carbonic acid 2.0 mM/l?

7. Using the table of factors, calculate the total CO_2 of a plasma specimen with a pH of 7.38 and a pCO_2 of 42 mm Hg.

8. Calculate the pCO_2 of a patient's plasma if the pH is 7.48 and the total CO_2 is 27.5 mM/l?

CHAPTER 10

RENAL CLEARANCE TEST CALCULATIONS

10.1 UREA CLEARANCE TEST

One of the first generally accepted tests of the overall function of the kidney was the urea clearance test. Urea undergoes both glomerular filtration and tubular reabsorption, and its clearance from the plasma is expressed as the milliliters of plasma that contain the amount of urea excreted by the kidneys each minute. Obviously the plasma is not completely cleared of urea because of the constant tubular reabsorption, but the common expression for clearance is the **milliliters of plasma cleared per minute**. Since the rate of urea reabsorption depends on the rate of reabsorption of water, we take this fact into consideration by using either the formula for **maximum clearance** (urine excretion rate 2.0 ml/min or more) or that for **standard clearance** (urine excretion rate up to 2.0 ml/min).

10.2 MAXIMUM CLEARANCE

Urea clearance tests are more significant when the urine excretion rate is 2.0 ml/min or more. Tubular reabsorption of urea has less influence on the test at this rate. Maximum clearance is calculated as follows:

$$C_{max} = \frac{U \times V}{S}$$

where C_{max} is expressed in ml/min; the concentration of urea in the urine (U) and serum (S) are both expressed in mg/100 ml, and V, the volume of urine, is expressed in ml/min. **The average normal maximum clearance is 75 ml/min**, and because there is also a standard urea clearance calculation, the clearances are often reported in per cent of the normal clearance.

Example

A patient's serum urea was 12 mg/100 ml, the urine urea was 222 mg/100 ml and the total urine volume for a 2 hour test period was 360 ml. The rate of urine excretion would be

$$\frac{360 \text{ ml}}{120 \text{ min}} = 3 \text{ ml/min}$$

Using the above equation:

$$C_{max} = \frac{U \times V}{S} = \frac{222 \times 3}{12} = \frac{666}{12} = 55.5 \text{ ml/min}$$

Therefore C_{max} = 55.5 ml/min or $\frac{55.5}{75}$ = 74 per cent of normal

10.3 STANDARD CLEARANCE

When the urine excretion rate is less than 2.0 ml/min, the normal urea clearance values drop below 75 ml/min. Since this decrease in clearance rate is approximately proportional to the square root of the urine excretion rate, the standard urea clearance is calculated as follows:

$$C_{std} = \frac{U \times \sqrt{V}}{S}$$

The average normal standard urea clearance is 54 ml/min.

Example

A patient's serum urea was 24 mg/100 ml, the urine urea 600 mg/100 ml and the total urine volume for a 2 hour test period was 180 ml. The rate of urine excretion would be

$$\frac{180}{120} = 1.5 \text{ ml/min}$$

Using the above equation:

$$C_{std} = \frac{U \times \sqrt{V}}{S} = \frac{600 \times \sqrt{1.5}}{24} = \frac{600 \times 1.23}{24} = \frac{735}{24} = 30.6 \text{ ml/min}$$

Therefore C_{std} = 30.6 ml/min or $\frac{30.6}{54}$ = 56.7 per cent of normal.

10.4 CREATININE CLEARANCE

In many laboratories the creatinine clearance test has replaced the urea clearance test as a measure of kidney function. Creatinine is predominantly excreted by glomerular filtration and therefore serves as an indicator of glomerular filtration rate. Since tubular reabsorption is minimal, we need not be concerned about maximum or standard clearances or per cent of normal clearance values. The calculation employs the simple relationship:

$$C = \frac{U \times V}{S}$$

Example

A patient's serum creatinine was 1.1 mg/100 ml, the urine creatinine 45 mg/100 ml and the urine volume for a 4 hour test period was 550 ml. The rate of urine excretion would be

$$\frac{550}{240} = 2.3 \text{ ml/min}$$

Using the above equation:

$$C = \frac{U \times V}{S} = \frac{45 \times 2.3}{1.1} = 94.2 \text{ ml/min}$$

Therefore the creatinine clearance is 94.2 ml/min, which may be compared to an average normal glomerular filtration rate of 125 ml/min for men and 115 ml/min for women.

To obtain meaningful comparisons of renal clearance rates for patients of various sizes and weights **the results must often be corrected for body surface.** This correction is obviously important in pediatrics. In adults the average body surface is taken as 1.73 square meters, and clearance test rates in both adults and children are corrected by the following formula:

$$C \times \frac{1.73}{A} = \text{ml/min corrected for body surface}$$

where A is the body surface of the patient in square meters. **Tables and nomograms** are available that yield body surface values in square meters when the patient's weight is expressed in kilograms (or pounds) and their height in centimeters (or feet and inches). If these aids are not available, the body surface may be calculated by a complex formula:

$$\text{Log A} = 0.425 \log W + 0.725 \log H - 2.144$$

where A is the body surface in square meters, W is the weight in kilograms and H is the height in centimeters.

Example

A patient weighs 120 pounds and is 5 feet 2 inches in height. The weight in kilograms = 120/2.2 = 54.5 kg and the height in centimeters = 62 inches × 2.54 cm/in = 157.6 cm.

From the above equation:

$$\text{Log A} = 0.425 \log W + 0.725 \log H - 2.144$$

$$= 0.425 \times \log 54.5 + 0.725 \times \log 157.6 - 2.144$$

$$= 0.425 \times 1.747 + 0.725 \times 2.198 - 2.144$$

$$= 0.744 + 1.592 - 2.144$$

$$= 2.336 - 2.144$$

$$= 0.192$$

$$A = \text{antilog } 0.192 = 1.555 \text{ square meters}$$

RENAL CLEARANCE TEST CALCULATIONS

The results of a creatinine clearance test on this patient were as follows: serum creatinine 1.2 mg/100 ml, urine creatinine 48 mg/100 ml and urine volume for a 4 hour test period was 500 ml. Calculate the creatinine clearance corrected for body surface.

From the equation:

$$C = \frac{U \times V}{S} = \frac{48 \times 2.08}{1.2} = 99.8 \text{ ml/min}$$

From the equation:

$$C \times \frac{1.73}{A} = \text{ml/min corrected for body surface}$$

$$99.8 \times \frac{1.73}{1.555} = 110.0 \text{ ml/min}$$

EXERCISES

1. Calculate the urea clearance and per cent of normal from the following data. Serum urea 15 mg/100 ml, urine urea 325 mg/100 ml and urine volume for a 2 hour test period was 300 ml.

2. A patient with glomerulonephritis excreted 120 ml of urine in a 2 hour test period. The urine contained 750 mg of urea per 100 ml and the serum 50 mg/100 ml. Calculate the clearance and per cent normal for this patient.

3. A patient's serum creatinine was 1.4 mg/100 ml, the urine creatinine 50 mg/100 ml and the urine volume for a 4 hour test period was 720 ml. Calculate the creatinine clearance.

4. A child weighed 25 pounds and was 2 feet tall. Calculate the body surface in square meters.

5. A child 4 feet 4 inches tall weighed 60 pounds. A creatinine clearance was run with a result of 95 ml/min (uncorrected). Correct the clearance for the child's body surface.

6. From the following data, calculate the creatinine clearance of the patient corrected for body surface. Serum creatinine was 0.9 mg/100 ml, urine creatinine 48 mg/100 ml, urine volume 2.2 ml/min, body weight 110 pounds and height 5 feet.

7. A large patient who was 6 feet 4 inches tall and weighed 240 pounds excreted 360 ml of urine in a 2 hour urea clearance test. The serum urea concentration was 12 mg/100 ml, and the urine urea was 400 mg/100 ml. Calculate the urea clearance corrected for body surface.

Appendix

TABLE OF FOUR

	0	1	2	3	4	5	6	7	8	9
1.0	.0000	.0043	.0086	.0128	.0170	.0212	.0253	.0294	.0334	.0374
1.1	.0414	.0453	.0492	.0531	.0569	.0607	.0645	.0682	.0719	.0755
1.2	.0792	.0828	.0864	.0899	.0934	.0969	.1004	.1038	.1072	.1106
1.3	.1139	.1173	.1206	.1239	.1271	.1303	.1335	.1367	.1399	.1430
1.4	.1461	.1492	.1523	.1553	.1584	.1614	.1644	.1673	.1703	.1732
1.5	.1761	.1790	.1818	.1847	.1875	.1903	.1931	.1959	.1987	.2014
1.6	.2041	.2068	.2095	.2122	.2148	.2175	.2201	.2227	.2253	.2279
1.7	.2304	.2330	.2355	.2380	.2405	.2430	.2455	.2480	.2504	.2529
1.8	.2553	.2577	.2601	.2625	.2648	.2672	.2695	.2718	.2742	.2765
1.9	.2788	.2810	.2833	.2856	.2878	.2900	.2923	.2945	.2967	.2989
2.0	.3010	.3032	.3054	.3075	.3096	.3118	.3139	.3160	.3181	.3201
2.1	.3222	.3243	.3263	.3284	.3304	.3324	.3345	.3365	.3385	.3404
2.2	.3424	.3444	.3464	.3483	.3502	.3522	.3541	.3560	.3579	.3598
2.3	.3617	.3636	.3655	.3674	.3692	.3711	.3729	.3747	.3766	.3784
2.4	.3802	.3820	.3838	.3856	.3874	.3892	.3909	.3927	.3945	.3962
2.5	.3979	.3997	.4014	.4031	.4048	.4065	.4082	.4099	.4116	.4133
2.6	.4150	.4166	.4183	.4200	.4216	.4232	.4249	.4265	.4281	.4298
2.7	.4314	.4330	.4346	.4362	.4378	.4393	.4409	.4425	.4440	.4456
2.8	.4472	.4487	.4502	.4518	.4533	.4548	.4564	.4579	.4594	.4609
2.9	.4624	.4639	.4654	.4669	.4683	.4698	.4713	.4728	.4742	.4757
3.0	.4771	.4786	.4800	.4814	.4829	.4843	.4857	.4871	.4886	.4900
3.1	.4914	.4928	.4942	.4955	.4969	.4983	.4997	.5011	.5024	.5038
3.2	.5051	.5065	.5079	.5092	.5105	.5119	.5132	.5145	.5159	.5172
3.3	.5185	.5198	.5211	.5224	.5237	.5250	.5263	.5276	.5289	.5302
3.4	.5315	.5328	.5340	.5353	.5366	.5378	.5391	.5403	.5416	.5428
3.5	.5441	.5453	.5465	.5478	.5490	.5502	.5514	.5527	.5539	.5551
3.6	.5563	.5575	.5587	.5599	.5611	.5623	.5635	.5647	.5658	.5670
3.7	.5682	.5694	.5705	.5717	.5729	.5740	.5752	.5763	.5775	.5786
3.8	.5798	.5809	.5821	.5832	.5843	.5855	.5866	.5877	.5888	.5899
3.9	.5911	.5922	.5933	.5944	.5955	.5966	.5977	.5988	.5999	.6010
4.0	.6021	.6031	.6042	.6053	.6064	.6075	.6085	.6096	.6107	.6117
4.1	.6128	.6138	.6149	.6160	.6170	.6180	.6191	.6201	.6212	.6222
4.2	.6232	.6243	.6253	.6263	.6274	.6284	.6294	.6304	.6314	.6325
4.3	.6335	.6345	.6355	.6365	.6375	.6385	.6395	.6405	.6415	.6425
4.4	.6435	.6444	.6454	.6464	.6474	.6484	.6493	.6503	.6513	.6522
4.5	.6532	.6542	.6551	.6561	.6571	.6580	.6590	.6599	.6609	.6618
4.6	.6628	.6637	.6646	.6656	.6665	.6675	.6684	.6693	.6702	.6712
4.7	.6721	.6730	.6739	.6749	.6758	.6767	.6776	.6785	.6794	.6803
4.8	.6812	.6821	.6830	.6839	.6848	.6857	.6866	.6875	.6884	.6893
4.9	.6902	.6911	.6920	.6928	.6937	.6946	.6955	.6964	.6972	.6981
5.0	.6990	.6998	.7007	.7016	.7024	.7033	.7042	.7050	.7059	.7067
5.1	.7076	.7084	.7093	.7101	.7110	.7118	.7126	.7135	.7143	.7152
5.2	.7160	.7168	.7177	.7185	.7193	.7202	.7210	.7218	.7226	.7235
5.3	.7243	.7251	.7259	.7267	.7275	.7284	.7292	.7300	.7308	.7316
5.4	.7324	.7332	.7340	.7348	.7356	.7364	.7372	.7380	.7388	.7396
5.5	.7404	.7412	.7419	.7427	.7435	.7443	.7451	.7459	.7466	.7474
5.6	.7482	.7490	.7497	.7505	.7513	.7520	.7528	.7536	.7543	.7551
5.7	.7559	.7566	.7574	.7582	.7589	.7597	.7604	.7612	.7619	.7627
5.8	.7634	.7642	.7649	.7657	.7664	.7672	.7679	.7686	.7694	.7701
5.9	.7709	.7716	.7723	.7731	.7738	.7745	.7752	.7760	.7767	.7774

APPENDIX

PLACE LOGARITHMS

	0	1	2	3	4	5	6	7	8	9
6.0	.7782	.7789	.7796	.7803	.7810	.7818	.7825	.7832	.7839	.7846
6.1	.7853	.7860	.7868	.7875	.7882	.7889	.7896	.7903	.7910	.7917
6.2	.7924	.7931	.7938	.7945	.7952	.7959	.7966	.7973	.7980	.7987
6.3	.7993	.8000	.8007	.8014	.8021	.8028	.8035	.8041	.8048	.8055
6.4	.8062	.8069	.8075	.8082	.8089	.8096	.8102	.8109	.8116	.8122
6.5	.8129	.8136	.8142	.8149	.8156	.8162	.8169	.8176	.8182	.8189
6.6	.8195	.8202	.8209	.8215	.8222	.8228	.8235	.8241	.8248	.8254
6.7	.8261	.8267	.8274	.8280	.8287	.8293	.8299	.8306	.8312	.8319
6.8	.8325	.8331	.8338	.8344	.8351	.8357	.8363	.8370	.8376	.8382
6.9	.8388	.8395	.8401	.8407	.8414	.8420	.8426	.8432	.8439	.8445
7.0	.8451	.8457	.8463	.8470	.8476	.8482	.8488	.8494	.8500	.8506
7.1	.8513	.8519	.8525	.8531	.8537	.8543	.8549	.8555	.8561	.8567
7.2	.8573	.8579	.8585	.8591	.8597	.8603	.8609	.8615	.8621	.8627
7.3	.8633	.8639	.8645	.8651	.8657	.8663	.8669	.8675	.8681	.8686
7.4	.8692	.8698	.8704	.8710	.8716	.8722	.8727	.8733	.8739	.8745
7.5	.8751	.8756	.8762	.8768	.8774	.8779	.8785	.8791	.8797	.8802
7.6	.8808	.8814	.8820	.8825	.8831	.8837	.8842	.8848	.8854	.8859
7.7	.8865	.8871	.8876	.8882	.8887	.8893	.8899	.8904	.8910	.8915
7.8	.8921	.8927	.8932	.8938	.8943	.8949	.8954	.8960	.8965	.8971
7.9	.8976	.8982	.8987	.8993	.8998	.9004	.9009	.9015	.9020	.9026
8.0	.9031	.9036	.9042	.9047	.9053	.9058	.9063	.9069	.9074	.9079
8.1	.9085	.9090	.9096	.9101	.9106	.9112	.9117	.9122	.9128	.9133
8.2	.9138	.9143	.9149	.9154	.9159	.9165	.9170	.9175	.9180	.9186
8.3	.9191	.9196	.9201	.9206	.9212	.9217	.9222	.9227	.9232	.9238
8.4	.9243	.9248	.9253	.9258	.9263	.9269	.9274	.9279	.9284	.9289
8.5	.9294	.9299	.9304	.9309	.9315	.9320	.9325	.9330	.9335	.9340
8.6	.9345	.9350	.9355	.9360	.9365	.9370	.9375	.9380	.9385	.9390
8.7	.9395	.9400	.9405	.9410	.9415	.9420	.9425	.9430	.9435	.9440
8.8	.9445	.9450	.9455	.9460	.9465	.9469	.9474	.9479	.9484	.9489
8.9	.9494	.9499	.9504	.9509	.9513	.9518	.9523	.9528	.9533	.9538
9.0	.9542	.9547	.9552	.9557	.9562	.9566	.9571	.9576	.9581	.9586
9.1	.9590	.9595	.9600	.9605	.9609	.9614	.9619	.9624	.9628	.9633
9.2	.9638	.9643	.9647	.9652	.9657	.9661	.9666	.9671	.9675	.9680
9.3	.9685	.9689	.9694	.9699	.9703	.9708	.9713	.9717	.9722	.9727
9.4	.9731	.9736	.9741	.9745	.9750	.9754	.9759	.9763	.9768	.9773
9.5	.9777	.9782	.9786	.9791	.9795	.9800	.9805	.9809	.9814	.9818
9.6	.9823	.9827	.9832	.9836	.9841	.9845	.9850	.9854	.9859	.9863
9.7	.9868	.9872	.9877	.9881	.9886	.9890	.9894	.9899	.9903	.9908
9.8	.9912	.9917	.9921	.9926	.9930	.9934	.9939	.9943	.9948	.9952
9.9	.9956	.9961	.9965	.9969	.9974	.9978	.9983	.9987	.9991	.9996

% TRANSMISSION — ABSORBANCE CONVERSION CHART

% T	A	% T	A	% T	A	% T	A
1	2.000	1.5	1.824	51	.2924	51.5	.2882
2	1.699	2.5	1.602	52	.2840	52.5	.2798
3	1.523	3.5	1.456	53	.2756	53.5	.2716
4	1.398	4.5	1.347	54	.2676	54.5	.2636
5	1.301	5.5	1.260	55	.2596	55.5	.2557
6	1.222	6.5	1.187	56	.2518	56.5	.2480
7	1.155	7.5	1.126	57	.2441	57.5	.2403
8	1.097	8.5	1.071	58	.2366	58.5	.2328
9	1.046	9.5	1.022	59	.2291	59.5	.2255
10	1.000	10.5	.979	60	.2218	60.5	.2182
11	.959	11.5	.939	61	.2147	61.5	.2111
12	.921	12.5	.903	62	.2076	62.5	.2041
13	.886	13.5	.870	63	.2007	63.5	.1973
14	.854	14.5	.838	64	.1939	64.5	.1905
15	.824	15.5	.810	65	.1871	65.5	.1838
16	.796	16.5	.782	66	.1805	66.5	.1772
17	.770	17.5	.757	67	.1739	67.5	.1707
18	.745	18.5	.733	68	.1675	68.5	.1643
19	.721	19.5	.710	69	.1612	69.5	.1580
20	.699	20.5	.688	70	.1549	70.5	.1518
21	.678	21.5	.668	71	.1487	71.5	.1457
22	.658	22.5	.648	72	.1427	72.5	.1397
23	.638	23.5	.629	73	.1367	73.5	.1337
24	.620	24.5	.611	74	.1308	74.5	.1278
25	.602	25.5	.594	75	.1249	75.5	.1221
26	.585	26.5	.577	76	.1192	76.5	.1163
27	.569	27.5	.561	77	.1135	77.5	.1107
28	.553	28.5	.545	78	.1079	78.5	.1051
29	.538	29.5	.530	79	.1024	79.5	.0996
30	.523	30.5	.516	80	.0969	80.5	.0942
31	.509	31.5	.502	81	.0915	81.5	.0888
32	.495	32.5	.488	82	.0862	82.5	.0835
33	.482	33.5	.475	83	.0809	83.5	.0783
34	.469	34.5	.462	84	.0757	84.5	.0731
35	.456	35.5	.450	85	.0706	85.5	.0680
36	.444	36.5	.438	86	.0655	86.5	.0630
37	.432	37.5	.426	87	.0605	87.5	.0580
38	.420	38.5	.414	88	.0555	88.5	.0531
39	.409	39.5	.403	89	.0505	89.5	.0482
40	.398	40.5	.392	90	.0458	90.5	.0434
41	.387	41.5	.382	91	.0410	91.5	.0386
42	.377	42.5	.372	92	.0362	92.5	.0339
43	.367	43.5	.362	93	.0315	93.5	.0292
44	.357	44.5	.352	94	.0269	94.5	.0246
45	.347	45.5	.342	95	.0223	95.5	.0200
46	.337	46.5	.332	96	.0177	96.5	.0155
47	.328	47.5	.323	97	.0132	97.5	.0110
48	.319	48.5	.314	98	.0088	98.5	.0066
49	.310	49.5	.305	99	.0044	99.5	.0022
50	.301	50.5	.297	100	.0000		.0000

TABLE OF ATOMIC WEIGHTS
(Based on Carbon-12)

	Symbol	Atomic No.	Atomic Weight		Symbol	Atomic No.	Atomic Weight
Actinium	Ac	89	227	Mercury	Hg	80	200.59
Aluminum	Al	13	26.9815	Molybdenum	Mo	42	95.94
Americium	Am	95	[243]*	Neodymium	Nd	60	144.24
Antimony	Sb	51	121.75	Neon	Ne	10	20.183
Argon	Ar	18	39.948	Neptunium	Np	93	[237]
Arsenic	As	33	74.9216	Nickel	Ni	28	58.71
Astatine	At	85	[210]	Niobium	Nb	41	92.906
Barium	Ba	56	137.34	Nitrogen	N	7	14.0067
Berkelium	Bk	97	[249]	Nobelium	No	102	[253]
Beryllium	Be	4	9.0122	Osmium	Os	76	190.2
Bismuth	Bi	83	208.980	Oxygen	O	8	15.9994
Boron	B	5	10.811	Palladium	Pd	46	106.4
Bromine	Br	35	79.909	Phosphorus	P	15	30.9738
Cadmium	Cd	48	112.40	Platinum	Pt	78	195.09
Calcium	Ca	20	40.08	Plutonium	Pu	94	[242]
Californium	Cf	98	[251]	Polonium	Po	84	210
Carbon	C	6	12.01115	Potassium	K	19	39.102
Cerium	Ce	58	140.12	Praseodymium	Pr	59	140.907
Cesium	Cs	55	132.905	Promethium	Pm	61	[145]
Chlorine	Cl	17	35.453	Protactinium	Pa	91	231
Chromium	Cr	24	51.996	Radium	Ra	88	226.05
Cobalt	Co	27	58.9332	Radon	Rn	86	222
Copper	Cu	29	63.54	Rhenium	Re	75	186.2
Curium	Cm	96	[247]	Rhodium	Rh	45	102.905
Dysprosium	Dy	66	162.50	Rubidium	Rb	37	85.47
Einsteinium	Es	99	[254]	Ruthenium	Ru	44	101.07
Erbium	Er	68	167.26	Samarium	Sm	62	150.35
Europium	Eu	63	151.96	Scandium	Sc	21	44.956
Fermium	Fm	100	[253]	Selenium	Se	34	78.96
Fluorine	F	9	18.9984	Silicon	Si	14	28.086
Francium	Fr	87	[223]	Silver	Ag	47	107.870
Gadolinium	Gd	64	157.25	Sodium	Na	11	22.9898
Gallium	Ga	31	69.72	Strontium	Sr	38	87.62
Germanium	Ge	32	72.59	Sulfur	S	16	32.064
Gold	Au	79	196.967	Tantalum	Ta	73	180.948
Hafnium	Hf	72	178.49	Technetium	Tc	43	[99]
Helium	He	2	4.0026	Tellurium	Te	52	127.60
Holmium	Ho	67	164.930	Terbium	Tb	65	158.924
Hydrogen	H	1	1.00797	Thallium	Tl	81	204.37
Indium	In	49	114.82	Thorium	Th	90	232.038
Iodine	I	53	126.9044	Thulium	Tm	69	168.934
Iridium	Ir	77	192.2	Tin	Sn	50	118.69
Iron	Fe	26	55.847	Titanium	Ti	22	47.90
Krypton	Kr	36	83.80	Tungsten	W	74	183.85
Lanthanum	La	57	138.91	Uranium	U	92	238.03
Lawrencium	Lw	103	[257]	Vanadium	V	23	50.942
Lead	Pb	82	207.19	Xenon	Xe	54	131.30
Lithium	Li	3	6.939	Ytterbium	Yb	70	173.04
Lutetium	Lu	71	174.97	Yttrium	Y	39	88.905
Magnesium	Mg	12	24.312	Zinc	Zn	30	65.37
Manganese	Mn	25	54.9380	Zirconium	Zr	40	91.22
Mendelevium	Md	101	[256]				

* A value given in brackets denotes the mass number of the longest-lived or best-known isotope.

ANSWERS TO EXERCISES

Chapter 1

1. $0.35
2. $100,000
3. 104°F
4. −112°F
5. 23.3°C
6a. 1500
 b. 600
 c. 5
7a. 2.2×10^6
 b. 4.6×10^4
 c. 1.51×10^5
 d. 3.0×10^{-4}
 e. 4.4×10^{-7}
 f. 5.0×10^{-9}
8a. 9.43×10^9
 b. 4.032×10^{12}
 c. 6.82×10^{-9}
 d. 27

Chapter 2

Section 2.3
A. 0.7404
B. 2.6345
C. $\overline{2}.7924$ or $8.7924 - 10$
D. 3.5110
E. $\overline{1}.1761$ or $9.1761 - 10$

Section 2.4
A. 20.78
B. 1515.5
C. 1.847
D. 0.03774
E. 0.2327

Section 2.5
A. 19.7
B. 317.8
C. 0.0032

Section 2.6
A. 4.84
B. 33.25
C. 503.9

Section 2.9
A. 478,400
B. 43,240
C. 9.28
D. log 6.8573
E. log $7.7160 - 10$

Additional Exercises
1a. 2
 b. $\overline{2}$ or $8 - 10$
 c. $\overline{5}$ or $5 - 10$
 d. 1
 e. 0
4a. 604
 b. 3.973
 c. 0.02782
 d. 97.25
 e. 0.00428
2a. 0.6625
 b. 0.6625
 c. 0.8920
 d. 0.1552
 e. 0.8451
5a. 2310
 b. 9406
 c. 29210
 d. 118.5
 e. 1.055
3a. 2.8273
 b. 0.1760
 c. 4.5315
 d. $\overline{3}.5315$ or $7.5315 - 10$
6a. 5420
 b. 2.132
 c. 3.486
 d. 5.9170
 e. $\overline{6}.3800$ or $4.3800 - 10$

APPENDIX

Chapter 3

Section 3.2
A. 585
B. 735
C. 630
D. 130

Section 3.4
A. 56.1
B. 53,300
C. 52,200
D. 12.68
E. 2.66

Section 3.5
A. 8.82
B. 2350
C. 221

Section 3.6
A. 11.68
B. 78.1
C. 4.49
D. 187,300
E. 18.66

Section 3.7
A. 13.45
B. 492
C. 48.6
D. 664,000
E. 8.12
F. 23.8
G. 0.2085
H. 2.55
I. 0.01298
J. 71.7

Section 3.8
A. 58.45 cm
B. 4.57 in
C. 8.87 ml
D. 324.0 mg
E. 54.5 kg
F. 110 lbs
G. 12.88 km
H. $X = 11.12$
I. $X = 0.75$
J. $X = 30$

Section 3.9
A. 826
B. 630
C. 79.0
D. 52,150
E. 11.5
F. 1680
G. 27.5
H. 46.5
I. 8.54
J. 0.0667
K. 154 lbs
L. $X = 38$
M. 241.3 km

Additional Exercises
1a. 2860
 b. 227.8
 c. 92,250
 d. 297,300
 e. 1,400,000

2a. 1760
 b. 67.25
 c. 114.4
 d. 22.46
 e. 570.5

3a. 980
 b. 2.28
 c. 23.5
 d. 0.2368
 e. 8.215

4a. 358 cm
 b. 91.44 cm
 c. 90.9 kg
 d. 59.1 kg
 e. 648 mg
 f. 80.95 mg
 g. 96.6 km
 h. 236.6 ml

5a. $X=40$
 b. 110.1 lbs
 c. 7.71 gr
 d. 8800 yds
 e. 33.83 oz

Chapter 4

Section 4.1
A. 57.5 gm

Section 4.2
A. 12.9 liters
B. 13.6
C. 4.46 gm

Section 4.3
A. 63%
B. 108.5 gm

Section 4.4
A. 60 gm
B. 17.55 μgm
C. 100 mM

Section 4.6
A. 29.4 gm
B. 0.145 N
C. 5 liters

Section 4.7
A. 200 ml
B. 825 ml
C. 45 ml
D. 10 mg/100 ml
E. 600 mg/100 ml

Section 4.8
A. 109.2 ml
Ba. 12.2 N
Bb. 36.9 N
Bc. 12.5 N

Additional Exercises
1. 50 gm
2. 1.020
3. 99 gm
4. Anhydrous
5. 0.098 gm
6. 46.8 gm
7. 2.72 ml
8. 580 ml
9. 266.7 ml
10. 29.2%
11. 250 ml
12. 104.2 mEq/l

APPENDIX

Chapter 5

1. Both x and y are variables while a and b are constants. As the graph of an equation for a straight line when y is plotted on the ordinate, x on the abscissa; the slope of the line is a and the intercept on the y axis is b.

2. x and y since x is usually plotted as the concentration of a standard solution or as time in hours and y is ordinarily plotted as absorbance readings or as concentration as in glucose tolerance curves.

3.

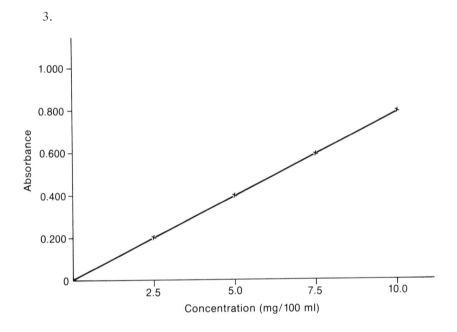

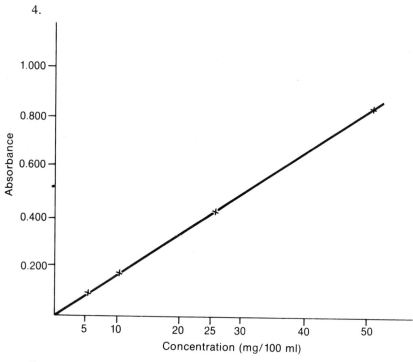

4.

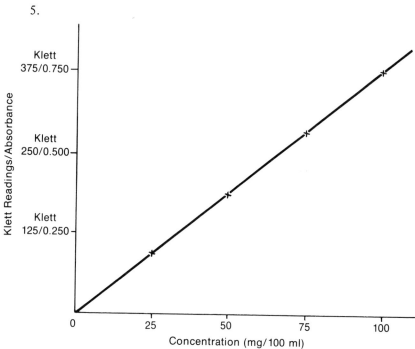

5.

APPENDIX

Chapter 6

Exercises
A. 4.25×10^3 M   B. 1.085

Additional Exercises
1. 18,200   3. 38.5 mg/100 ml   4. 1.2 mg/100 ml
2. 0.545

5. No, Beer's law is not followed at the two highest concentrations. Slight variation at 4 mg%, considerable at 5 mg%.

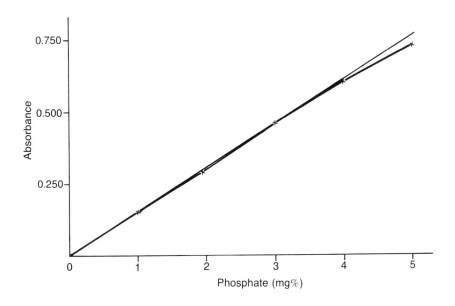

6a. 4.32 mg%
b. 3.38 mg%
c. 3.30 mg%
d. 4.46 mg%

7a. 125.6 mEq/l
b. 141.2 mEq/l
c. 134.4 mEq/l
d. 135.8 mEq/l

7.

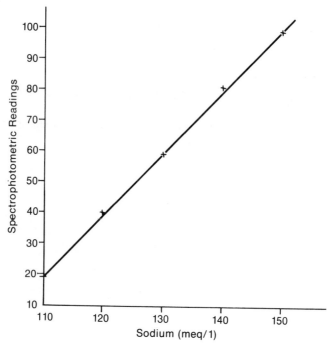

Chapter 7

Exercises

A. 1.49 mEq/l

B. Males S = 0.216 C. V. 3.54%
    Females S = 0.134 C. V. 2.34%

Additional Exercises

1.
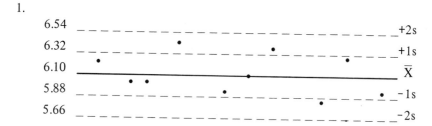

2.

```
12.71 _____•_____ +2s
12.52 _____•_____•_____ +1s
12.33 _____•__•_____•_____•__••_____ X̄
12.14 ___•_____•_____•_•_____ -1s
11.95 _____•_____•_____•_____ -2s
```

3.

```
102.52 _____•_____ +2s
100.76 _____•_____•____ +1s
 99.00 _____•_____•_____ X̄
 97.24 ____•_____•_____•_____ -1s
                             •                 
 95.48 _____•_____ -2s
```

Chapter 8

Exercises

1. 0.0003
2. 0.095
3. 1×10^{-4}; 4
4. 3.4
5. 0.014; 1.8; 1.85
6. 7.51
7. 3.16×10^{-4}

Chapter 9

Exercises

A. pH 7.10
B. pH 5.73
C. pK_a 6.10
D. pH 4.49

Additional Exercises

1. pH 7.23
2. 44.5 mm Hg
3. 41.2 mm Hg
4. 23.5 mM/l
5. pH 7.55
6. 25.2 mM/l
7. 25.4 mM/l
8. 36.6 mm Hg

Chapter 10

Exercises

1. 54.2 ml/min.; 72.3%
2. 15 ml/min.; 27.8%
3. 107.2 ml/min
4. 0.394 square meters
5. 95 ml/min
6. 141 ml/min
7. 72 ml/min

Index

Absorbance, 55, 56, 57, 59, 62, 63, 65, 66, 67, 68
Accuracy of a method, 71
Acetate buffer, 80, 81
   preparation of, 82, 83
Acetic acid, dissociation of, 80, 82
Acid-base buffer calculations, 84-88
Alternate division – multiplication method, 28, 36
Answers to exercises, 101-108
Antilogarithms, 12
Approximation of an answer, 6, 7
Arithmetical common sense, 1
Atomic weights, table of, 100

Beer's law, 65, 67
   equation for, 65
Bicarbonate-carbonic acid buffer, 84, 85
Blood, buffers, 84-88
   glucose values, statistical table of, 72
   pH, 84, 85, 87
Body surface area, 92, 93
Buffers, 80-88
   acid-base, 84-88
   bicarbonate-carbonic acid, 84, 85
   equations for, 80, 81
   Henderson-Hasselbalch equation for, 81, 84, 85, 86
   maximum capacity of, 81
   pK_a of, 81, 85
   preparation of, 82, 83

Carbon dioxide, partial pressure of, 85, 86, 87
   total content of, 86, 87

Carbonic acid, dissociation constant of, 81, 82
Centigrade (Celsius) scale, 2
   conversion to Fahrenheit, 2, 3
Characteristic, of a logarithm, 9
Circular slide rule, 33-38
   diagram of, 35
Clearance, standard, 89, 90
Coefficient of variation, 72, 73
   equation for, 73
Combination multiplication and division problems, 27, 36
Concentration of solutions, 41-54
Conversion, problems in, 31, 37, 51
   table of units of, 31
Coordinates, choice of, for graphs, 56, 57, 58
Creatinine clearance, 91, 92, 93
Curves, glucose tolerance, 58
   graphs of, 58
   standard, 58-63
   graphs of, 58-64

Decimal point, 6, 24, 30
   correct setting of, 6, 7, 24, 25
   in square root problems, 30
Deviation, standard, 72, 73, 74, 75
   equation for, 73
Dilution, of solutions, 48, 68, 69
   of a specimen, 50, 68
   of a standard, 50, 69
Division, 5, 14, 25, 35
   of numbers on a slide rule, 21
   using logarithms, 14
      powers of ten, 5
      slide rule, 25, 35

INDEX

Establishing the correct decimal point, 6, 7, 24, 30

Fahrenheit scale, 2
   conversion to centigrade, 2, 3

Glomerular filtration rate, 91
Glucose tolerance curve, 58
Gram-equivalent weight, 47
Graph paper, proper use of, 56, 57, 58
Graphs, 56, 57, 58, 60, 61, 62, 63
   choice of coordinates for, 56, 57, 58
   equation of, 55
   of glucose tolerance curve, 58
   of standard curves, 58–64
   preparation of, 55–64

Handling numbers, 4
   as powers of ten, 4, 5
Henderson-Hasselbalch equation, 81, 84, 85, 86
Hydrates, 44
Hydrogen ion concentration (pH), 76–79
   calculation of, 76, 78

Indicator, on a slide rule, 20, 34

Logarithm(s), 9–19
   characteristic of, 9
   mantissa of, 10
   of a number, 11
      expressed as powers of ten, 17
   table of, 97–98

Mantissa, of a logarithm, 10
Maximum buffer capacity, 81
Maximum clearance, 89, 90
Mean value, 71, 72, 73, 74, 75
   equation for, 71
Method, accuracy of, 71
   precision of, 71
   reliability of, 71
Metric system, 2
Micromoles, 46
Milliequivalents, 52
Millimoles, 46
Molal solutions, 47

Molar extinction coefficient, 65
   calculation of, 66, 67
Molar solutions, 45
Mole, 45
Multiplication, 4, 5, 13, 23, 34
   and division problems, 14, 27
   using logarithms, 13
   using powers of ten, 4, 5
   using slide rule, 23, 25, 34

Normal distribution curve, 74
Normal solutions, 47
Numbers, 4, 11
   corresponding to a logarithm, 12
   handling of, 4
   logarithms of, 11
      expressed as powers of ten, 17
   powers of, 15
   roots of, 16, 29, 37
   setting of, on a slide rule, 21, 22, 34

Partial pressure, of CO_2 (pCO_2), 85, 86, 87
   in relation to Henderson-Hasselbalch equation, 85, 86
   table for calculation of, 87
Per cent transmission (% T), 59, 60, 61, 65, 67
Percentage solutions, 41
   preparation of, 42, 43
   types of, 41
pH (hydrogen ion concentration), 76–79
   approximation of, 77, 78
   calculation of, 77
pK of a buffer, 81
Powers, of numbers, 15
   of ten, 4
      negative, 4, 5
      positive, 4, 5
Precision of a method, 71
Preparation, of buffers, 82, 83
   of graphs, 55–64
   of solutions, 42, 43, 47, 49
   of standard curves, 58–63
Proportion problems, 31, 32, 37

Quality control, chart used in, 74, 75
   statistics in, 71–75

Range of values, 7, 72, 73
Reliability of a method, 71

INDEX

Renal clearance test, calculations for, 89–94
  correction for body surface area in, 92, 93
Roots of numbers, 16, 29, 37

Slide rule, 20–40
  alternate division-multiplication method, 28, 36
  circular, 33–38
  conversion and proportion problems on, 31, 37
  description of, 20, 33, 34
  division of numbers on, 21, 34
  multiplication on, 23, 25, 34
  scale of, 20, 21, 33, 34
  setting the decimal point on, 24
  setting of numbers on, 21, 22, 34
  square and square roots on, 29, 37
Solubility coefficient, for CO_2, 86
Solutions, concentration of, 41–54
  dilution of, 48, 68, 69
  molal, 47
  molar, 45
  normal, 47
  percentage, 41
    preparation of, 42, 43
    types of, 41
  preparation of, 42, 43, 47, 49
Specific gravity, 43
Specimen, dilution of, 50, 68
Spectrophotometric calculations, 65–70

Squares and square roots, 16, 29, 37
Standard, clearance, 89, 90
  curves, 58–63
  deviation, 72, 73, 74, 75
    equation for, 73
  dilution of, 50, 69
Strong acid, 76
Surface area, of body, 92, 93

Total CO_2 content, 86, 87
Transmission-absorbance conversion chart, 99

Urea clearance test, 89, 90
  maximum clearance of, 89, 90
  standard clearance of, 89, 90

Values, range of, 7, 72, 73
Variation, coefficient of, 72, 73

Water, dissociation of, 76
Water of hydration 44
Weak acid, 76
Weight, gram-equivalent, 47

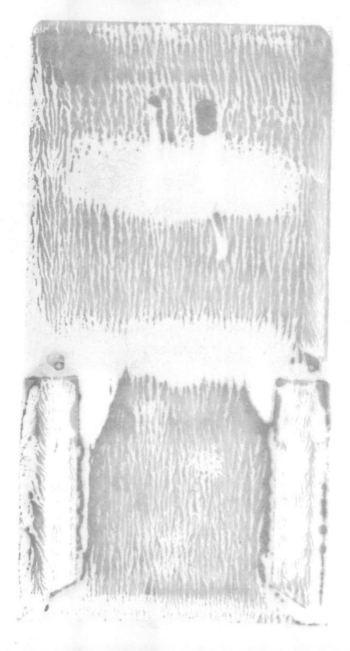